Timbering and Mining

By W.H. Storms

with an introduction by Kerby Jackson

This work contains material that was originally published in 1909.

This publication was created and published for the public benefit, utilizing public funding and is within the Public Domain.

This edition is reprinted for educational purposes and in accordance with all applicable Federal Laws.

Introduction Copyright 2014 by Kerby Jackson

Introduction

It has been nearly one hundred and ten years since McGraw-Hill released it's important publication "Timbering and Mining: A Treatise on Practical American Methods". First released in London in 1909, this important volume has now been out of print for over a century and has been unavailable to the mining community since those days, with the exception of expensive original collector's copies and poorly produced digital editions.

It has often been said that *"gold is where you find it"*, but even beginning prospectors understand that their chances for finding something of value in the earth or in the streams of the Golden West are dramatically increased by going back to those places where gold and other minerals were once mined by our forerunners. Despite this, much of the contemporary information on local mining history that is currently available is mostly a result of mere local folklore and persistent rumors of major strikes, the details and facts of which, have long been distorted. Long gone are the old timers and with them, the days of first hand knowledge of the mines of the area and how they operated. Also long gone are most of their notes, their assay reports, their mine maps and personal scrapbooks, along with most of the surveys and reports that were performed for them by private and government geologists. Even published books such as this one are often retired to the local landfill or backyard burn pile by the descendents of those old timers and disappear at an alarming rate. Despite the fact that we live in the so-called "Information Age" where information is supposedly only the push of a button on a keyboard away, true insight into mining properties remains illusive and hard to come by, even to those of us who seek out this sort of information as if our lives depend upon it. Without this type of information readily available to the average independent miner, there is little hope that our metal mining industry will ever recover.

This important volume and others like it, are being presented in their entirety again, in the hope that the average prospector will no longer stumble through the overgrown hills and the tailing strewn creeks without being well informed enough to have a chance to succeed at his ventures.

                                                                       Kerby Jackson
                                                                        Josephine County, Oregon
                                                                        July 2014

CONTENTS

CHAP.		PAGE
	Introduction	v
I.	Timber used in Mining	1
II.	Preservation of Timbers	8
III.	Drifting and Drift Sets	16
	Driving lagging in drifting	24
IV.	Driving in Running Ground	28
	Breast boards	29
V.	Structural Steel in Mine Workings	34
VI.	Timbering Drift Gravel Mines in California	38
	Breasting posts and caps	38
VII.	Shafts	42
	Location, kind and size of shafts. Substantial timbering usually necessary	42
	Prospecting shafts in good ground and those requiring close timbering. Cribs and single-compartment shafts	46
	Size and division of working shafts. Drainage	49
	Positions of temporary hoisting plant and permanent plant	52
	Size of shaft compartments	54
	The collar	55
	Hanging-bolts	58
	The cross-head in sinking shafts with a bucket	58
	Stull methods in many small mines of the Cobalt district, Ontario	62
VIII.	Bucket Dumping	64
	Methods at vertical shafts	64
	Methods at inclined shafts	69
IX.	Framing Shaft Timbers	73
	Framing for vertical shafts	75
	The divider and spliced wall plate	80
	The template	85
	Handling timbers in shafts	85
	Some useful knots	88
	Placing timbers of shaft sets in position	89
	Placing timbers in inclined shafts and lining the sets	93
	Difficulties in sinking through running ground	96
	Sinking through running or loose ground	97
	Combination vertical and inclined shafts	102
X.	Bearers in Shafts	105
	Fenders for the protection of shaft timbers when blasting	111
	The cable mat	112

CONTENTS

CHAP.		PAGE
	Extension tracks for sinking	113
	Inverted rails	114
	Sinking-ladders	115
	Inclines in hard rock	115
	Shaft repairing	116
XI.	Position and Direction of Drill Holes in Shaft Sinking	117
XII.	Cutting and Timbering Stations at Shafts	119
	Ore pockets beneath the levels	120
	Types of stations at inclined and vertical shafts	125
	Drainage by means of skips and tanks	125
XIII.	Mining Large Ore Bodies by the Open-cut or "Glory-hole" System	133
	The churn-drill or jumper	141
	The steam shovel in open cuts	144
XIV.	The Overhand and Underhand Methods of Stoping Veins	148
XV.	Unusual Methods of Stoping Enforced by Scarcity of Timber	154
XVI.	Stoping in Flat or Low-lying Veins	158
XVII.	Raises for Connection of Levels	161
XVIII.	Stoping and Filling in a Mine having Weak Walls	166
XIX.	The Caving System Practised at the Pewabic Mine, Iron Mountain, Michigan	172
XX.	Stoping in Swelling Ground	178
	Main gangways driven in the country rock	179
	Connecting levels when stoping veins of moderate width	184
XXI.	Stoping Large Ore Bodies	187
	Introduction of the square-set by Philip Deidesheimer on the Comstock	188
	Usefulness of the square-set in extracting large ore bodies	191
	Mistakes made in using the square-set	193
XXII.	Framing Square-set Timbers	197
	Construction of the square-set in the stope	201
XXIII.	Modifications of the Square-set System in California Mines	205
	Construction of chutes in square-sets	211
XXIV.	Stoping Large Ore Bodies by the Block System at Broken Hill, New South Wales	213
	Underground open-cut and sloping-stope systems, at Broken Hill	218
XXV.	Mining at the Homestake, Lead, South Dakota	229
	Stoping without timbers at the Homestake	238
	Usefulness of mine models	243
XXVI.	Working Dangerous Ground in the Kimberley Diamond Mines	247
XXVII.	The Delprat Method of Stoping without Timbers	253
XXVIII.	Head-frames	255

Chapter I

TIMBER USED IN MINING

Miners throughout the Western United States prefer, as a rule, those kinds of timber commonly known as spruce and Oregon pine, the latter being really what is technically called Douglas spruce, though it is also variously known as Douglas fir, yellow fir, or red fir, according to its color. In addition to these two kinds of timber, yellow pine, sugar pine, pinon (bull) pine, several varieties of fir, and, in a few localities, oak, are used for the support of mine workings.

In the Coast Range of California, redwood, though usually considered unsuited to the purpose, has been extensively employed in timbering quicksilver mines, particularly at New Almaden, in Santa Clara County. Redwood, though brittle, is enduring, and has been found to answer the purpose very well indeed, where the pressure has not been too great.

On the desert, yuccas, juniper, cottonwood, scrub pine, old railroad ties, in fact, almost any sort of wood available, has been used by prospectors in timbering their workings, though in some of these mines we have observed that the timbers had died out, and fallen down, while the ground still stood without support. It is not always possible for miners to obtain the timber they prefer, and therefore they naturally turn to whatever kind of timber may be available and least expensive in their locality.

In 1904 an attempt was made by the writer to ascertain from various sources what kind of timber had been found most satisfactory in mine workings, and with that idea in view a number of mine superintendents in California were requested to give the result of their experience in this connection. The replies disclosed the fact that all were not of one mind on the subject. The information was desired primarily as the basis of a paper to be read before the California Miners' Association. To this end the following questions were sent to each of the several gentlemen whose replies are given below. The questions covered the following points:

1. Kind of timber used; whether yellow or sugar pine, spruce, fir, Oregon pine, etc.

2. Condition when placed in the mine — seasoned or not.

3. Position in the mine with reference to excavation, that is, whether in stope, drift or shaft, and relation of air currents, as there seems to be considerable difference in the enduring qualities of timber, depending on its location — whether in still air or in a current of considerable velocity; and also whether the air current is pure and fresh or whether heavily laden with carbon dioxide and the other foul gases peculiar to mines.

4. Humidity of air; is the stope damp? Does fungus form rapidly on the timbers, or is it comparatively dry and free from these growths?

5. How long should a timber, say 20 to 30 in. diameter, endure under average conditions, in a well-ventilated drift in a damp mine? (*a*) spruce; (*b*) yellow pine; (*c*) sugar pine; (*d*) fir; (*e*) redwood.

6. What is the best timber for general use in mines? Is there a material difference in the various kinds of timber, of equally good condition, that is, well-seasoned sticks?

To the questions submitted, Mr. George W. Starr, managing director of the Empire mines at Grass Valley, California, replied as follows:

No. 1. Yellow pine and spruce.

No. 2. Used after being cut eight months.

No. 3. We use yellow pine in stopes and in places where not much life of timber is required. Spruce is used in drifts and main tunnels.

No. 4. Stopes are usually damp and fungus forms after six months.

No. 5. (*a*) In wet ground where water is trickling over timber constantly, spruce will last indefinitely, but in damp ground three to five years; (*b*) yellow pine, about two years.

No. 6. Spruce.

The Empire mines are opened on small veins with low dip in granite, and present unusual conditions, being very unlike the mines of the Mother Lode of California.

Mr. John Ross, Jr., a California mine manager of many years' experience, replied to the questions as follows:

No. 1. Oak, spruce, yellow pine, Douglas spruce (commonly known as Oregon pine), bull (pinon) pine, sugar pine, native fir.

Oak is rarely used because of its scarcity, but I know a shaft that was sunk in 1859–60 and kept open for seven years, part of the shaft being timbered with oak. The mine then lay idle for twenty years, the shaft being full of water. It was reopened in 1887, and has been kept open until the present time [1904]. The oak timbers stood until 1901, and at that time the wall plates and sills were in perfect condition, and would not have been removed then, but for the fact that the pressure on the end timbers — or legs — had been great enough to crush the ends a little, making them too short to be used again. The shaft is an incline, and a little water was constantly trickling over the timbers — enough to keep them thoroughly wet all the time. Under the same conditions a few sets of bull (pinon) pine in the same shaft stood for forty years, while spruce, yellow pine, Oregon pine and fir had to be changed several times, the timber lasting in the order given, its life in the shaft being from two to eight years, where the timbers were always wet. In other parts of the shaft, where the timbers were wet and dry alternately, the life of the timber ranged from two to three years. This shaft, however, is an exceptionally destructive one to timbers, being sunk in the vein fissure and in the gouge where the pressure on the timbers was not equally distributed. In justice to the other members of the timber family, it is but fair to say that the oak and bull pine were used in the narrowest part of the vein, where the pressure was not so great as where the fissure was wider.

No. 2. The condition of the timbers going into the mine varies from green, fresh from the axe, to thoroughly seasoned — that is, timber that has been cut two years. Where the timbers going into the mine are sound, and they are placed where the water trickles over them, there is practically no difference in the life of a seasoned timber or a green one; but in a dry or damp drift, the seasoned timber is preferable, partly on account of less cost in handling and putting in place; and furthermore it is not so quickly attacked by either fungus or dry rot. I have seen green timber placed in a damp drift attacked by fungus within two weeks of the time it was put in place; but this is uncommon.

No. 3. The position of timbers in the mine relative to air currents, and the purity of the air, has more to do with its life, in my opinion, than anything else. In making a connection between two shafts 1000 ft. apart, work was commenced from each shaft, the air being forced down each shaft to the face of each drift, giving plenty of good air for workmen. The work was not rushed, taking about a year's time. During all this time the timbers stood well, showing little sign of either dry rot or fungus growth. When the connection was made the air came through the drift with considerable velocity, making one shaft a down-cast, and the other an up-cast. Near the downcast shaft the drift and stope were 30 ft. wide, and heavily timbered. These timbers stood for six years with practically no repairs. Between this place and the up-cast shaft the drift was opened out to a width of 20 to 30 ft. in a number of places in stoping, the drift being kept as straight as possible, and through this most of the air came. It was quickly noted that the further we got from the down-cast shaft, the worse the condition of the timbers became, they being attacked with fungus growth and dry rot. It was also noticed that where the drift had been widened by stoping, timbers out of the direct line of the air current were not so quickly attacked, and experience later showed that they would last twice as long. The life of timbers that were in direct line with the air current was from one to two years, and when not fairly in the current from two to four years. It was also noted that the timbers in the up-cast shaft, above the point of connection, soon became seriously affected, necessitating considerable repair work. As the property became more extensively developed, this experience was duplicated in three different levels.

Subsequently, in opening a drift in a caved portion of one of these connecting levels, where the drift proper had been retimbered three times before it caved, it was found that the drift timbers were little better than punk, while the stope timbers that came down with the caving of the drift were as sound as when placed in position. The stope timbers had been out of the direct line of the air current, and many of these stope timbers had been in place three to four years longer than those in the drift.

No. 4. Warm, moist air, in a mine that has but one shaft, is far more destructive on timber than dry and comparatively cool air. In the latter instance, the timber will last at least twice as long.

No. 5. (a) Spruce, about six years; (b) yellow pine, about five years; (c) sugar pine, about four years; (d) fir, about two years; (e) redwood — never used any.

No. 6. For all-around work, spruce.

At one time a comparative test was made in the Wildman mine, at Sutter Creek, Cal., in a raise where the ground was very heavy. Oregon pine, 12 × 12 in., was used on one side, and 14-in. round native yellow pine on the other. In one year the round, yellow pine stick was badly bent and twisted, and the legs were forced into the wall-plate three inches, but there was still strength enough in the stick to hold the ground, while the 12 × 12 in. Oregon pine was shattered into slivers, and could bear no further pressure. None of the timbers used in this work showed any rot or fungus growth.

Mr. G. McM. Ross, a mine manager and superintendent of large experience in the West, contributed the following in reply to the questions:

No. 1. I have used many kinds of timber in various conditions, from green, wet wood, fresh from the sawmill pond, to old telegraph poles, railroad ties, sage brush, and the juniper of Nevada; also the fiber of the tall cactus of Arizona.

The above answers the second question also.

No. 3. Any timber will last longer in a stope or drift that is well ventilated with fresh, pure air, than where there is a limited amount of fresh air or none at all.

No. 4. Timber of any kind will last longer underground when wet, or when there is sufficient moisture in the ground or air to keep it damp. Fungus will not form on timbers where there is a good circulation of pure air, but forms rapidly where there is little or no air.

No. 5. It is impossible to figure on an average condition. We have seen large timbers crushed into slivers within a few days underground, and again have seen timbers of the same size perfectly sound after being underground for forty years. Spruce, yellow pine, fir or cedar, in a well-ventilated damp mine, where there is no excessive pressure, should last for fifty years. Sugar pine, which is the best timber we have, is now too valuable for mining work in heavy ground. Timber that will best resist end-thrust is the most serviceable for mining work.

Mr. E. Hampton, who at the time these questions were sent out was superintendent of the Oneida mine, near Jackson, Amador County, California, responded as follows:

The timbers used in the Oneida mine are yellow, bull and sugar pine, spruce, fir and cedar (native timbers); 50 per cent. is bull pine. Some Oregon pine is also used. There are only a few places in the mine where timber of any kind has a chance for a natural life — that is, to stand long enough to resist decay; the ground being very heavy, there is nothing left to tell the tale, but where conditions exist that make it possible to determine the life of a timber, I find that bull pine timbers, 20 in. in diameter, put in place in a well-ventilated dry drift, four years ago, are now rotten clear through, without any perceptible fungus growth. I have observed that timber will decay in the Oneida mine as readily in still air as in a current of air; that fungus forms on green as well as on dry timbers, but that it forms more rapidly on dry timbers, and that green timbers last longer than dry timbers; that moisture in abundance, where pressure does not exist, has a tendency to increase the life of timbers indefinitely regardless of ventilation, especially spruce. We cleaned out a drift in 1903 that had been run in 1870 and timbered with spruce, and found that those timbers that were not broken were as solid as when first cut.

For all-round mining purposes I consider best: first, spruce; second, yellow pine; third, bull pine, and, in this order, sugar pine, fir, and cedar. In heavy ground there is but little use for the last three, but I believe that cedar will resist decay longer than any other timber. Redwood I know nothing about. In February, 1896, the Oneida shaft was started and timbered with 12 × 12 in. Oregon pine. In July, 1904, it had to be retimbered from the collar to 50 ft. below. The timbers were all rotten; changes of season and of atmosphere must have been the causes, as gas from the mine workings below never reached the timbers in the shaft, it being a down-cast. At 75 ft. below the collar of the shaft moisture is a fixed quantity, and from that point downward there is no sign of decay. Oregon timber is good all-round timber, but not equal to California spruce. With pressure it splits too easily. Perhaps if the dealers would send to the mines the same kind and quality of timber that they send to the University for testing purposes, we might change our opinion.

Mr. Frank F. Weber, assistant superintendent of the New Almaden quicksilver mines, Santa Clara County, California, in response to the questions, offered the following:

During the last few years the timber used in our main stopes and gangways has been Oregon pine, although in the early history of the mine redwood timbers, either hewn or sawed, were universally employed, and the old workings accessible afford good opportunities for observing the efficiency of this kind of timber under varying conditions. At the time of going into the mine the Oregon timber had been cut probably not longer than three or four months, so that while it can hardly be classed as well seasoned, it is not as green as that used in some localities, where timber, abundant in the vicinity, is cut, framed, and ready for use almost immediately. Generally speaking, I believe that the condition most unfavorable to the life of any mine timber is one of warmth with an abundance of moisture. In stopes where this condition is met with, the effect on pine as observed here is rapid decay, the timber being attacked by fungus, occasionally, within from four to six months. Redwood under the same condition does not seem to be attacked so readily by the fungus, and its life is considerably prolonged.

In making a comparison of the relative enduring qualities of redwood and pine under favorable conditions — in an unusually cold and damp tunnel — the following facts were observed: 10×10 in. redwood timbers, put in place over thirty-five years ago, while being sound at the center, showed a decided softening near the surface, although they are still serving their purpose satisfactorily. In the stopes requiring considerable timber to support the ground, Oregon pine (Douglas spruce) has been employed, and is found to be the most satisfactory in every way. Especially is this true where the square-set system is used, for in taking weight this kind of timber will stand a surprising amount of buckling and twisting before absolute failure. Redwood, on the other hand, will not support weight nearly as well. It is brittle, will split and give way with little warning — sometimes so quickly that repairs are often prevented — while pine shows at what point repairs are needed and allows ample time in which to do such work.

With regard to the effect of carbon-dioxide gas on timber, I do not believe that it has any noticeable effect one way or another. It is met with in considerable quantities in portions of this mine, and any particular effect would be noticed. On the other hand, foul gases arising from a pile of decaying timber will in all probability cause a favorable condition for like decay in adjacent timbers. It would be a hard matter to give an accurate estimate as to how long a stick of timber would endure, even under favorable conditions. However, I should judge that seasoned redwood, in a wet drift or shaft, such as that heretofore mentioned, and coming in contact with a current of fresh air, should last for all practical purposes from thirty to forty years; Oregon pine, under like conditions, from fifteen to twenty-five years.

In my opinion, the best timber for all-round mining work is well seasoned pine of the better varieties found throughout the Northwest. As a rule, it is a strong, safe timber to work with, and its life is sufficiently long for all practical purposes. Wherever possible, the class of timber employed should be that which best suits the conditions met with, and an ideal state of affairs would be to use redwood in all shafts and main gangways, and timber the stopes and drifts of lesser importance with pine. The price of redwood (which in this locality, New Almaden, is nearly twice that of pine) often prohibits its use.

It is evident from the several opinions of the gentlemen quoted above that each speaks according to his own experience, all practically agreeing that for general — in fact, for most — purposes, spruce is best.

The statement of Mr. Frank F. Weber, that carbon-dioxide has no appreciably destructive effect upon the timbering at New Almaden is of interest, as some portions of that mine are noted for the large amount of this gas issuing from crevices in the rocks. He recognizes, however, that carbon-dioxide due to decomposition of timbers in the mine has a noticeable effect in promoting decay of new timbers placed in the mine workings. It has been suggested, in this connection, that possibly there may be an important difference in these gases — between that emanating from the rocks and that due to rotting of timber — and that possibly the latter may consist in part at least of hydrocarbons.

Chapter II

PRESERVATION OF TIMBERS

In view of the rapid deterioration of timbers placed in most mines, not wholly or at all due to heavy pressure, engineers have for years past experimented in an effort to discover some means of lessening this rapid decay of timbers, and several methods have been found which preserve timber for a greater or less length of time. Among those who have made an exhaustive study of methods of preserving timber for use in mines and elsewhere, are the large railroad companies, and more particularly the Forest Service of the Government. Gifford Pinchot, Forester, has caused to be issued, as Circular 111 of the Forest Service, a pamphlet entitled "Prolonging the Life of Mine Timbers," written by John M. Nelson, Jr., forest assistant. From this little work the following notes have been abstracted:

"*Peeling the Timber.* — Experiments have shown that peeled timber is superior in durability to unpeeled timber. The space between the bark and the wood especially favors the development of wood-destroying fungi and is a breeding place for many forms of insect life. When, after placement in the mines, the bark begins to flake off, the timber has already begun to decay. The cost of peeling timber before it goes into the mine ranges from 20 cents to 50 cents per ton of wood, according to local conditions and the kind of timber.

"*Seasoning the Timber.* — Seasoning or drying gives mining timber greater strength and durability. A stick of wet timber has only about one-half the strength of a similar stick absolutely dry. Though it is not practicable for mining companies to hold their timber until it is absolutely air dry, peeled timber will dry out sufficiently in a few months to gain in both strength and durability. From two to four months is necessary for proper seasoning. If a mining company handles its own timber from the woods to the mines, the saving in freight made possible by peeling and seasoning can readily be estimated. Labor is the principal factor in the cost of peeling, while the cost of seasoning must be represented by the loss of interest on the capital invested in the timber during the seasoning period. However, these additional items of expense are more than offset by a maximum reduction in freight of from 30 to 40 per cent., and by the far better condition of the timber with regard to both its life at the mines and the readiness with which it will take preservative treatment. The peeling of timber at the mines has been unsatisfactory and expensive, because of the limited amount of yard room and the accumulation of bark. The following considerations favor peeling in the woods: (1) The saving in the cost of freight due to peeling and seasoning; (2) the saving of yard room at the mines; and (3) the prevention of fungus disease and insect attack by early peeling.

"*Treating the Timber.* — Peeling and seasoning mine timber unquestionably increase its durability. However, in order to prolong its life to the fullest extent, a preservative treatment is necessary. Impregnated wood resists decay because the preservative is antiseptic and excludes the moisture necessary for fungus growth. Timber used in the mines was treated with a variety of preservatives under several methods of application. Both green and seasoned timbers were treated to determine both the relative value of the treatments and the best method of handling preparatory to treatment. If treated at all, the timber must be peeled.

"*Brush Treatments.* — Brush treatments with both creosote and carbolineum were applied in two coats to the Pennsylvania and Southern pines. A large flat brush and a kettle of the hot preservative are all that is required for this treatment. A very small amount of the preserving fluid suffices, but the cost of application in proportion to the results obtained is considerable. For small individual operators who can not afford the cost of a large plant, brush treatments are feasible and economical. The disadvantages of brush treatments are: (1) The difficulty of completely covering the timber and filling all checks and cracks. (2) The very slight penetration secured. The subsequent checking or opening of the timber may often allow disease to pass through the shallow exterior band into the untreated interior wood.

"*Open-tank Treatments.* — Pitch pine and loblolly pine have been most successfully treated with both creosote (dead oil of coal tar) and a six per cent. solution of zinc chloride by the open-tank process.

"*Description of Tank.* — The experimental open tank was, for the most part, constructed from old material already in the possession of the company. A section of an old boiler 34 in. in diameter and 13 ft. in length was set vertically in the ground to a depth of 5 ft. This tank had a double bottom, separated by a space of one foot. Between the two bottoms a coil of 1-in. pipe 20 ft. in length carrying a steam pressure of 110 lbs. per square inch furnished the heating surface necessary to give the preservative fluid a maximum temperature of 240° F. This coil was connected by a 1-in. pipe to a 10-in. steam main 75 ft. distant. The timbers, which were placed vertically in the tank, were immersed by attaching a circular weight to their lower ends. The timbers were lowered into and hoisted from the tank by means of a small hand derrick with a swinging arm.

"*Description of the Treatment.* — The open-tank treatment as given in this experiment was briefly as follows: Green, partially seasoned, and thoroughly seasoned timber was lowered into the tank and immersed in creosote, or in a zinc chloride or salt solution, at a temperature of from 90° to 120° F. The temperature of the creosote was raised by the coils to from 212° to 220° F., and that of the zinc chloride or the salt solution to about 212° F. In no case, however, was the temperature allowed to go above 240° F., for fear of injuring the fiber of the timber and so decreasing its strength. When this hot bath was over the steam was turned off and the timber was allowed to stand until the liquid cooled to a temperature of from 170° to 100° F. The periods of heat and of cooling were varied for each kind of timber and for each stage of its seasoning. The time required for the cooling operation, which depended largely upon the temperature of the atmosphere, was usually from 3 to 12 hours. For the whole treatment the time varied from 6 to 20 hours.

"*Theory of the Open-tank Process.* — The theory of the open-tank process may be given in a few words. The heat of the preservative expands and expels a portion of the air and water contained in the cellular and intercellular spaces of the wood tissue, and as the preservative cools there is a contraction and condensation of the air and water which remain. To destroy the partial vacuum thus formed, the liquid is forced by atmospheric pressure into the cellular and intercellular spaces, a process aided, of course, by capillary attraction. In point of fact, therefore, the hot bath merely prepares the wood for absorbing the preservative, and the actual impregnation follows as the preservative cools. The ease and effectiveness with which timber can be treated by this process depend upon the kind of wood and its degree of dryness. In one species the structure of the wood tissues may effectually resist, and in another may greatly favor, the expulsion of air and water during the hot bath; in seasoned timber air, and in green timber water, is the chief element to be removed before the wood can be impregnated, and since air may be expelled much more easily than water, seasoned timber is the more successfully treated.

"*Possibilities and Regulations of the Treatment.* — Loblolly and pitch pine, among the more important mining timbers of the anthracite region, have been treated by the open-tank process with particular success. By simply immersing the timber first in hot and then in cold preservative fluids, a penetration of from 1 in. in green timber to from 4 to 5 in. in seasoned timber has easily been secured. Aqueous solutions of zinc chloride and common salt have been absorbed with as much ease as creosote. In timbers which have a considerable proportion of heartwood the line of demarcation separating heartwood and sapwood is frequently also the line separating the treated and the untreated wood. In past experiments with the open-tank process, the heartwood has not been penetrated to a great depth, though this may be accomplished hereafter. The sapwood of chestnut and red oak has been treated with a fair degree of success, but with extremely little penetration of the heartwood. With absolutely green timber it has been a question of obtaining the greatest possible impregnation in the entire period of treatment (20 hours). For green timber the period of heat or preparation for treatment has been increased from its usual length (7 hours) to 18 hours, and this timber has been given two separate entire treatments on successive days without any improvement over the standard treatment of 20 hours. The treatment of loblolly and pitch pine is regulated by the proportion of heartwood and sapwood contained and the degree of seasoning reached. With a stick of pine, absorption is more easily controlled by varying the duration of the cold bath than by varying that of the hot bath. The duration of the hot bath necessary to prepare any form or kind of timber for impregnation is exceedingly variable. Seasoned timber, however, absorbs the cooling preservative with a fair degree of regularity down to a temperature of 120° F.

"*Summary of the Open-tank Treatment.* — (a) Loblolly and pitch pine can be successfully and economically treated by simple immersion in successive hot and cold baths in an open tank at

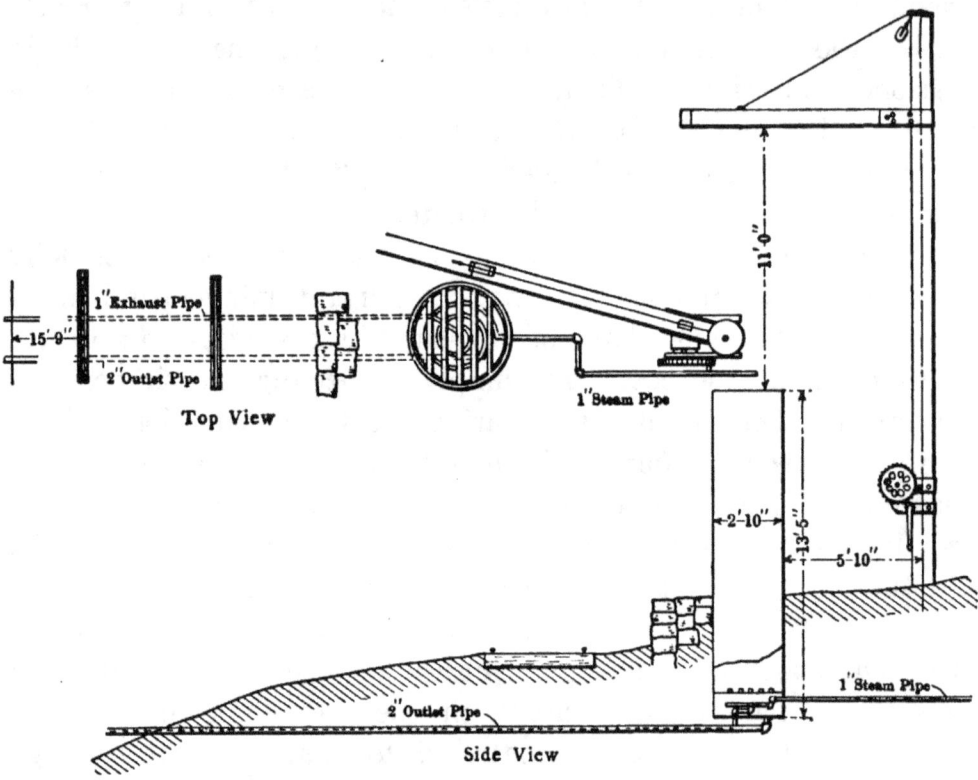

Fig. 1. — Plant for Treatment of Mine Timbers

a cost of about 11 cents per cubic foot. (b) Green timber is treated with far more difficulty than seasoned timber. (c) The difference in weight of green timber before and after treatment is by no means indicative of the amount of the preservative absorbed. The simple application of the hot liquid to green timber slightly reduces its weight and yields no penetration. The same application to seasoned timber slightly increases its weight and gives a slight penetration. Green timber after treatment may show a penetration of 1 in. without an increase in weight. (d) Heartwood of both loblolly pine and pitch pine is penetrated with far more difficulty than is the sapwood of the same species. This is especially the case with pitch pine, which clearly shows after treatment a distinct division between the treated sapwood and the untreated heartwood. (e) Experiments indicate that for pine timbers of the same degree of dryness, or containing equal proportions of heartwood and sapwood, impregnation can be regulated by increasing or decreasing the duration of the cooling bath.

"*Cylinder Treatment.* — Loblolly pine gangway timber was treated for experimental purposes in a closed cylinder under pressure. One portion of this timber was treated with 12 lbs. of creosote per cubic foot, and the remainder with a 5 or 6 per cent. solution of zinc chloride.

"*Method of Treatment.* — The timber to be treated was loaded on trucks and drawn into a 6-ft. steel cylinder by means of a wire cable. The doors of the cylinder were closed, and steam was turned in at the required pressure for from 4 to 6 hours. When the steam had been allowed to escape and the condensed water in the cylinder to run off, a vacuum of 22 in. was applied. During this process steam was passed through the heating coils within the cylinder. The preservative fluid was then run into the cylinder and pressure was applied until the desired absorption was attained.

"*Comparative Cost of Open-tank and Cylinder Treatments.* — The method of treatment in a closed cylinder under pressure is effective but expensive. Here, as in the previous treatments, the cost of the preservative is a large item, but the cost of application is far greater by the cylinder process than by the others. The saving secured by the open-tank method is due to: (1) The omission of the steam, vacuum, and pressure features of the cylinder process and the elimination of the expensive machinery necessary for those stages of the treatment. (2) The light construction of the tank, allowed by the lack of strain on the walls (3) The small amount of labor required in the operation of an open-tank plant, due to the simplicity of its construction and method of applying the preservative. (4) The fact that the cost of construction and maintenance of an open-tank plant is less than one-fifth that of a cylinder plant of equal capacity.

"*Results Derived from Experiments.* — Though results so far derived from actual experiments do not cover all classes of mine timber under all conditions, they show that it will unquestionably pay mining companies to peel their round timber, to season it for a few months, and to treat it thoroughly with some good preservative. For pitch pine and loblolly pine, the open-tank process with creosote has proved an efficient and economical method of treatment. The preservative value of zinc chloride for mining purposes is yet to be determined. Gangway timbers treated with creosote by the cylinder process are standing well. Because of its cost, however, this form of treatment should not be considered unless, in comparison with the far less expensive open-tank method, it gives universally better results. Timbers treated by the brush method with creosote and carbolineum have so far effectively resisted decay. Because of the very simple method of application, brush treatment may prove advantageous for small consumers, or where the timber is in great danger of being broken by excessive crushes.

"As a direct result of these co-operative experiments, the company is considering the advisability of treating their mine timber on a more extensive scale. Plans have been drawn up for the construction and erection of a commercial open-tank wood-preserving plant at one of its collieries. This plant will have a daily capacity of about thirty sets of gangway timber (800 cu. ft.) and will be large enough to treat all timber at this colliery except that which is broken or worn out. Creosote or a solution of zinc chloride, or both, will be the preservative fluids used, although the plant is designed for the use of any preservative which may prove efficient.

"*A Timber Policy for Preservative Treatment.* — If a mining company has proved by actual experiment that timber preservation is practical and economical, it should be in a position to carry it out. To do this timber cannot be rushed directly from the woods to the mines; there must be time for preparing it for treatment and for treating it. This means the storage of a reserve supply of felled timber at one or more points. To insure a regular supply of timber for their present and future needs, mining companies should purchase and operate tracts of timber land. For such an investment to be permanent, the logging must be carefully and economically done, and the forest protected from fire and managed on sound principles. The timber should be peeled immediately upon being felled in the woods, and allowed to season while waiting on cars for shipment. In this way freight charges on bark and a portion of the water present in green wood will be saved and the timber will be rendered more resistant to decay. In many cases the time consumed before and during transportation may be enough to season the timber sufficiently to prepare it for preservative treatment on its arrival at the mines. A careful and thorough inspection of all timber is strongly recommended. It would be poor economy to apply expensive preservative treatments to defective material. Timber cut from land owned by the mining company should be inspected in the woods or at the point of shipment, to avoid unnecessary freight charges. Timber shipped to the mines by outside parties should be just as carefully inspected. At present, timber is sometimes accepted in such condition that it is doubtful whether its service in the mines would pay for the cost of setting it, exclusive of the cost of the timber. No matter how critical the timber situation may be, it is believed that the policy of accepting everything is a poor one."

Chapter III

DRIFTING AND DRIFT SETS

Having discussed at length the various kinds of timber used in Western metal mines generally, and the means employed to enable this timber to resist premature decay, we will now take up the practical methods of timbering mines, beginning with the simplest forms of mine openings — drifts and cross-cuts. While it is true that drifts and cross-cuts are essentially the simplest form of mine workings, the problem of sustaining the ground through which such excavations pass is not always simple, by any means, for in these workings are encountered slips, which may cause an unanticipated fall of rock with possibly serious and sometimes fatal consequences. There are zones of fracture to be passed through wherein the ground may be heavy and require extreme care in handling, and in the placing of timber-sets to secure it; running ground is another bane of the miner's life, and this also requires special methods and experienced men. Then there is swelling ground, not so dangerous, usually, as either heavy, running or caving ground, but most difficult to support in such manner that it can be held without removal of the timber at frequent intervals. Swelling ground is the most expensive ground, usually, that the metal miner is required to handle, and yet even this can be so managed as to give the minimum of trouble.

In timbering drifts, as in every sort of mine workings, there are certain underlying mechanical principles which must be observed if success in preventing collapse of the workings is to be attained. As no two mines are exactly alike, and as different kinds of rock often require differing kinds of artificial support when an excavation has been made in them, the miner must necessarily be versatile in his timber schemes, and in the manner of applying them. The accompanying sketches show some of the simplest methods of placing drift timbers: Fig. 2 illustrates a method of placing timbers in ground which stands fairly well, but which may slack and drop upon exposure. The posts are set upright; no sill is provided beneath the posts, the ground being firm and dry; but this is a matter for mature judgment for if the rock on the floor of the drift be of such character that it is firm and solid when dry, but becomes soft and crumbling if wet, sills should be placed, in anticipation of the drift encountering water further in, which, flowing out would wet and soften the ground near the mouth of the tunnel and cause the sets to settle. All of this may be avoided by timely placing of the sills, as shown in Fig. 3. Tracks may be laid directly upon the sills with ties intermediately between the sets. Whether wet or dry, drains should be provided for the passage of any water that may

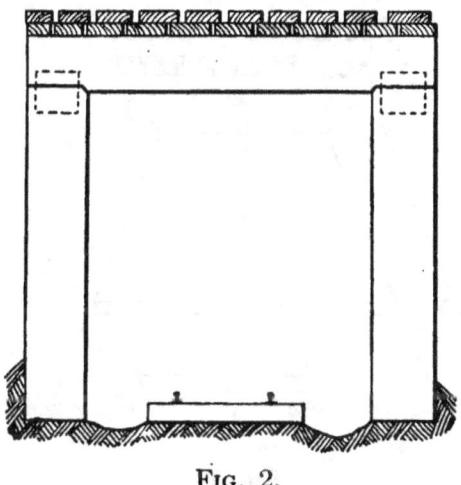

Fig. 2.

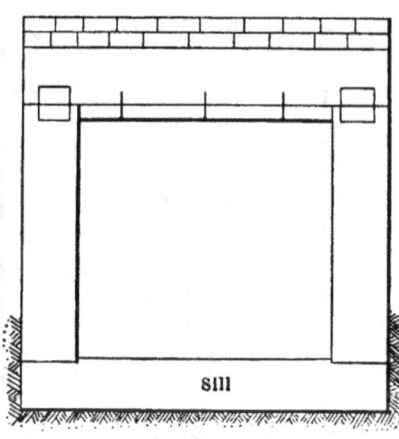

Fig. 3

be subsequently encountered. The drains should be cut on both sides of the drift or beneath the center. The matter of drainage is a most variable one, some mines making little or no water, others having far more than their share of it.

While the sketches given (Figs. 2, 3 and 4) show drift sets placed in an upright position, it is customary to give the posts an outward inclination, so that the posts are farther apart at the bottom than at the top. This is shown in Fig. 5. Upright or square drift sets are well enough suited to easy ground where there is no great weight to support, and where the ground after being cut stands well for several hours or days. Lagging may then be driven with little or no trouble, the ends being kept pointed upward by the use of wedge-shaped blocks. A set of lagging having thus been placed, those of the next set in advance are inserted beneath the forward ends of those of the previous set, and driven ahead as the work advances. The particular methods of driving lagging will be fully described and illustrated later.

Fig. 4 shows a method of framing a square drift set which is frequently seen in mines, but which has absolutely nothing to recommend it, except that in the event of heavy side pressure the posts cannot easily be thrust inward, being prevented by the deep shoulder of the cap. This style of framing results in the loss of about half the strength of the cap, and the only thing accomplished by it can be better done in another way.

A very simple drift set is illustrated in Fig. 3, which may be adapted either to the upright posts or to those spread as shown

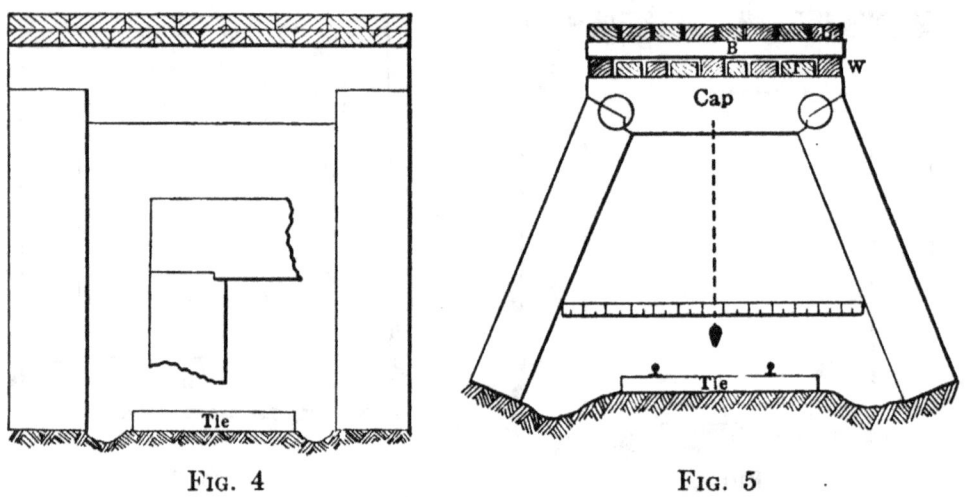

Fig. 4   Fig. 5

in Fig. 5. The set consists of a cap and two posts, neither of which are framed, being simply sawed off at a right angle, or at such angle as the spread of the posts requires. To prevent the posts from being crowded together, a plank two or three inches in thickness is spiked to the under side of the cap. This plank, being exactly the length the posts should be apart at the top, makes the sets uniform, and affords the necessary resistance to side pressure.

In Fig. 5 is shown a good method of centering sets: it is merely an aid to the eye. Of course, sets may be placed under the direction of an engineer using a transit, but this is a refinement in adjustment usually dispensed with by miners. The set may be driven to either side by means of wedges until the plummet line exactly coincides with the center mark on the staff, as shown.

In Fig. 4, in the center of the sketch, is shown a simple method of framing the post and cap in drifts sets, much in vogue in some camps. It may be described as a sort of double shoulder, and is designed on the theory that by its use neither cap nor post is likely to split under pressure: any scheme in timber framing which will accomplish this is well worthy of a trial. The "spiked plank" illustrated in Fig. 3 is one of the best of all the methods employed to secure the full strength of the timbers used.

Referring again to Fig. 5: This is a method of framing and placing drift sets that has no superior in any kind of ground. It is the method of framing and placing timbers most commonly employed on the Mother Lode of California, and is successfully used in the heaviest and worst swelling ground on the Lode. By successful is meant relatively successful as compared with other methods. No other scheme of framing or placing of timbers in heavy ground is known which affords better opportunities for holding the ground, or lasts longer under the unusually bad conditions often found there. All ground on the Mother Lode of California is not heavy, nor is it all bad. There is some very good standing ground there. Most of the mines, for instance, at Angels in Calaveras County, are in good ground. Occasionally, serpentine or talc is encountered, and then trouble sometimes comes quickly and often, but on the whole Calaveras County mines are not notoriously bad; but in Amador, the next county to the north, many of the leading mines have ground which none but those familiar with it can appreciate. Not only is this ground often heavy — a thick gouge-like mass of crushed slate with stringers of quartz and calcite, difficult to hold because of the dead weight, but often on exposure it swells, and exerts an irresistible pressure on the timbers. The heavy gouges may often be cut and timbered without great difficulty when the ground is dry, but should it become wet, then troubles multiply and danger increases, for the ground may run. The method of handling running ground will be treated later.

In most of the Mother Lode mines the workings are chiefly in the vein itself. Particularly is this the case with the workings of the earlier years. In many instances where a drift is run on the vein, whether on solid quartz or on gouge, the adjacent slate, which very often forms at least one wall and sometimes both, may be solid and firm, dipping with the vein, or near it, and showing no disposition when freshly broken to give the least trouble. A miner strange to the ground would consider 12-in. timber strong enough for work in such material, but the Mother Lode miner has learned by expensive experience that he dare not trust this seemingly kindly ground, so he puts in sets, disposed about as in Fig. 5, giving the legs a greater or less spread according to his ideas gained by experience. The main members of these are 24 in. and are not infrequently 30 in. or more in diameter. Until a few years ago, the custom was to place these sets 4 to 5 ft. apart (center to center) and lag them with heavy spiling, top and sides, driven closely together. In many instances, not only was the lagging quickly bent, twisted and broken, but unless relieved the posts were crowded in at the foot, or, being unable to move inward, due to rock obstruction, they would bend, split and break — often within two or three weeks of the time they were first placed.

Experience in time taught the mine superintendents that close lagging was not the scheme best suited to the conditions found on the Mother Lode, no matter what they might be elsewhere. Gradually they learned that the best results were obtained when heavy posts and caps were employed, and the lagging placed with considerable space between them, so that the swelling ground might crowd through the open spaces, and thus in some measure relieve the irresistible pressure. Later it was found to be of still further advantage to not only leave liberal spaces between the lagging, but to employ men to cut away the soft ground through these open spaces and to shovel the accumulation into cars from time to time and thus keep the main members in place. Keeping up swelling ground is always expensive, but long experience has proved that it is economy to treat it as here described — employing men to cut away and relieve the ground, replacing an occasional broken lagging, and thus maintaining the main members of the set in place.

In some of the drift mines, as well as quartz mines, of California, swelling ground is encountered. In the former it is the bed rock that swells. Where this is the case, either in a quartz or a drift mine, the bottom of the drift must be cut out from time to time and the track readjusted to grade. The legs of the timber sets must be given an unusually broad spread — even greater than is indicated in Fig. 5, as this generally gives more satisfactory results than where they are more nearly upright. In some instances, as at one place in the Oneida mine, near Jackson, Amador County, California, the swelling ground was so troublesome and such a constant source of expense that a drift was run through solid country rock of the foot-wall, around this piece of bad ground. Soon after this, the old drift was filled completely by the material crowded in from bottom and sides. We have seen old workings in some of the mines of Amador County which had been reopened by new drifts, where the only sign that mining had ever been done there was the old timbers, crushed and broken, that were embedded in the soft gouge-like material that had completely and compactly filled the old drift.

Thus far we have gone on the supposition that the ground through which we have been driving is fairly good standing ground, not giving unusual trouble in either cutting or timbering. In fact, the work could be carried far enough in advance of timber sets to admit of the sets being placed in position without extraordinary precautions. Often, however, conditions are not so fortunate, and the miner must be protected overhead and on either side as the work advances. With proper arrangements, even under conditions such as here anticipated, the work may advance with perfect safety and with rapidity. Loose ground, running ground, and that which is likely to spawl off and drop without warning, like the so-called "nigger heads" in serpentine, and any other ground that is in any way dangerous, may be worked and securely timbered by employing proper methods. We will now consider the methods of placing the timbers, driving lagging, the usefulness of the "false set," of breast boards, and other extraordinary methods of timbering drifts.

The methods of timbering drifts heretofore described are also applicable to inclines of low slope. Not infrequently miners drive tunnels on a vein having a low angle of dip. Mine workings of this character may be safely timbered in the same manner as those run at or near a level. In these inclined tunnels the dip of the vein sometimes makes radical changes, becoming steeper or flatter. This does not necessitate any change in the style of timbering, unless the vein plunges down so steeply that shaft sets are required, that the foot wall may also be sustained. In any event the miner will always bear in mind one principle — the posts must be set perpendicular to the roof of the incline (not vertical). As the angle of dip increases the posts are set forward at the top at a correspondingly increased angle to meet the downward pressure directly. This is somewhat different from the method of placing stulls (which will be dealt with later), but the posts must be set as above explained in order that the pressure may be evenly distributed, and not permit the timbers to "ride" either backward or forward. To still further insure against this, the timber sets, in all types of drift and tunnel, either horizontal or inclined, are held in place and kept from moving in either direction toward each other by timbers extending from set to set, and so disposed as to catch both post and cap at the corners. These timbers are called "sprags" in Ameri-

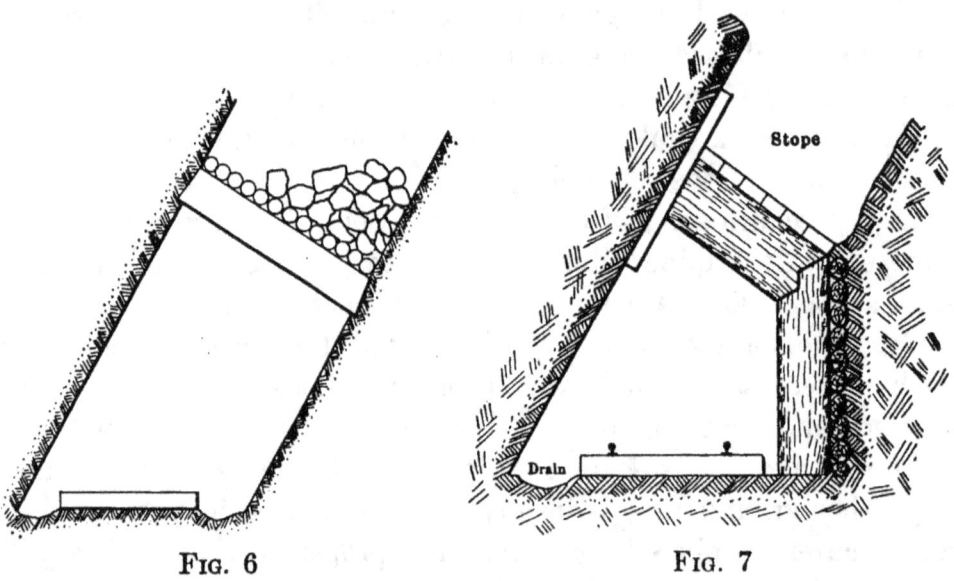

Fig. 6   Fig. 7

can practice. Sprags are also used in other systems of timbering, which will be described later. A drift, flat or inclined, timbered without the sets being properly spragged, will collapse upon the slightest movement of any set out of the perpendicular. And if one set goes down from this cause it is likely to so weaken others adjacent to it that the entire drift may be wrecked by the falling down or "jack-knifing" of sets.

In some ground it is considered desirable to place sprags both top and bottom of the sets — always at the top, and near the bottom when the footing of the posts is considered insecure. If the ground under the posts be particularly soft or slick (talc, serpentine or clay), it is better to employ sills upon which the posts may rest, thus rendering the sets far more secure. Occasionally a drift is run on a vein in which the ground stands so well that no timber sets such as described are necessary at all, stulls only being put in place, usually 8 to 9 ft. above the floor of the drift, upon which lagging may be laid, and waste rock for filling dumped upon the lagging after the ore has been removed. (See Fig. 6.) In other cases another means of support is employed, as where the distance between walls is considered too great or the character of the walls too treacherous to render simple stulls sufficiently secure. This idea is illustrated in Fig. 7. Both of these latter properly belong under the head of methods of stoping, and will be more fully considered later.

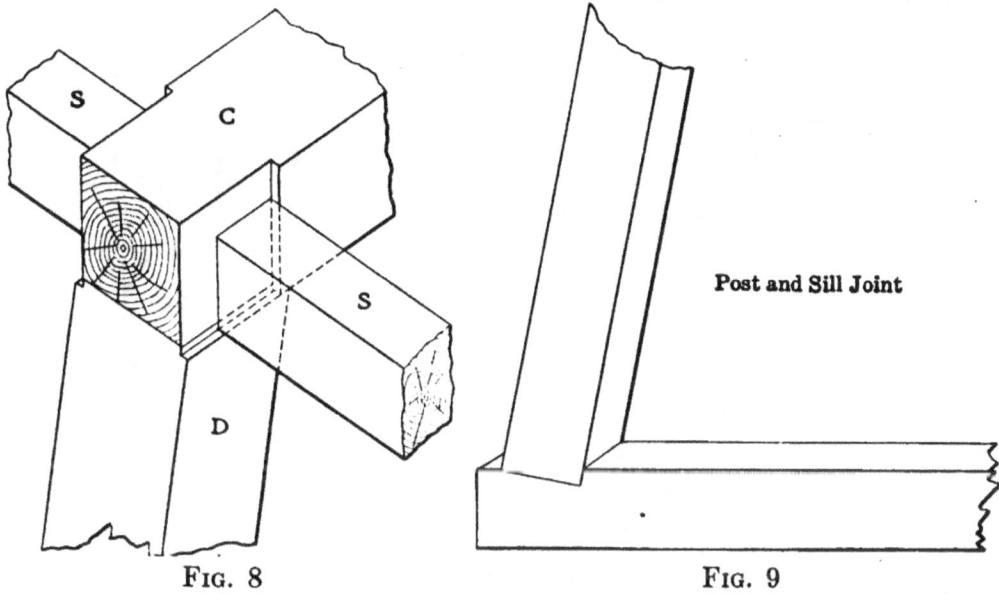

Fig. 8   Fig. 9

Fig. 8 shows the relative positions of post, cap and sprag in a drift set. Although the timbers here represented are square, round timbers may be framed in precisely the same manner. It will be observed that the sprag is so placed as to hold both cap and post in position. Some miners use spikes to secure sprags in place, but the use of spikes and nails is not to be recommended for this purpose.

Fig. 9 illustrates a good method of framing the sill so that the post sets squarely upon it, the foot of the post requiring no framing or bevel at all. It not only permits the post to have a good square footing on the sill, but the inside shoulder formed by the angle dap prevents the foot of the post from being easily crowded into the drift by side pressure.

Driving Lagging in Drifting

In driving drifts, cross-cuts and adits it is often imperative to support the ground by means of lagging driven ahead of the workmen, and as close to the face of the advancing cutting as possible. In ground of this sort it is not infrequently necessary to support the sides as well as the top of the working place. Where the ground stands fairly well, the timbers may be placed and the lagging driven at the convenience of the miners, often the work of excavating being carried 8 to 10 ft., or even more, in advance of the last set of timbers. By cutting the back of the drift high, and beveling off the upper edge of each piece of

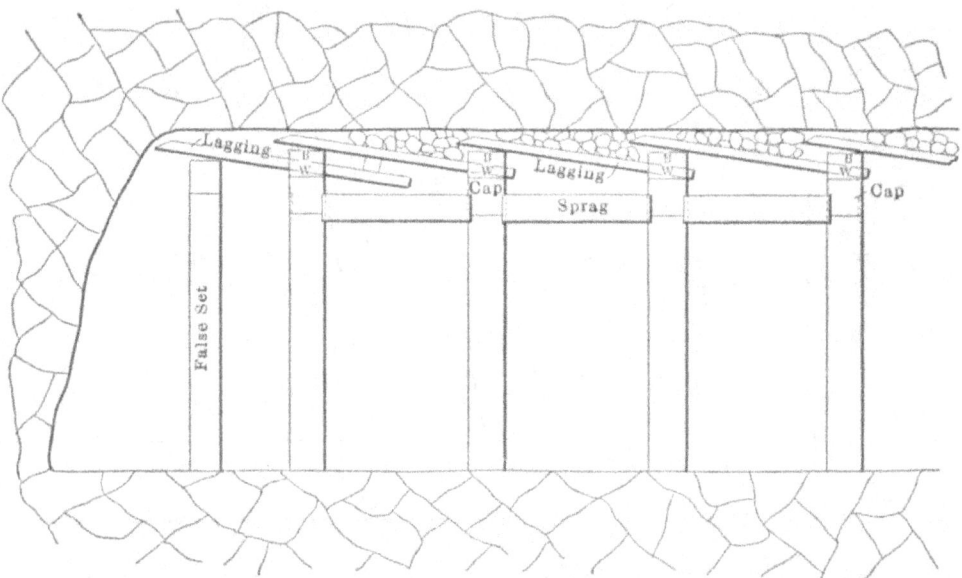

Fig. 10. — The Usual Drift Set with False Set

lagging, these — 5 to 8 in. wide and usually from 2 to 3 in. thick — are driven ahead, the forward ends being inserted beneath the ends of the last set driven to place. The forward ends are kept well upward by means of blocks, as hereafter explained. However, where the ground will not stand for an hour, or even a few minutes, the lagging must be driven forward inch by inch as the work of removing the ground progresses. When the ground requires this prompt and constant support, lagging cannot readily be driven in the manner above described; then a more elaborate scheme is adopted — that of wedge-shaped blocks and the "bridge." The method of advancing a drift by means of the bridge is indicated in Fig. 10. In the forward end of the drift,

DRIFTING AND DRIFT SETS 25

of which the sketch is a vertical longitudinal section, will be seen a set of timbers differing somewhat from the others. It has no bridge, and the posts of this set are slightly higher than those of the other sets. This is called a "false set," and is a most useful arrangement in advancing through unusually bad ground. When a set has been completed, and the work of excavation is continued, the workmen are protected and the drift rendered secure by inserting the forward end of the lagging under the bridge of the last set in place. The bridge B will be seen above each set, being separated from the cap in each set by the wedge-shaped block W. One of the blocks is placed at each end of the cap, and generally one in the center, to prevent the bridge from bending under the weight of the ground.

The lagging, having the upper edge beveled off, as shown in the sketch, is inserted between the bridge and the cap and is driven forward as the work progresses, the ends being kept pointed well upward by means of a block, shown in the last set, which is retained in position by the upward pressure of the lagging, it being occasionally necessary to force it into proper place by means of a few taps of a hammer. When the lagging has been driven ahead about $2\frac{1}{2}$ ft. beyond the center of the last set, the pressure of the heavy ground may be so great as to threaten to force the lagging down by bending it, or even causing it to break. When this is the case the false set is put in place, the posts being first set up and temporarily held in position while the cap is placed on the posts and the set wedged as firmly as possible against the sides of the drift. This false set is of such height as to catch the forward ends of the half-driven lagging, which may now rest upon the cap of the false set. When the work of excavation has again been advanced to a point sufficiently far to admit the placing of another regular set, the timbers are placed in position, the miners working in safety under the protection of the top lagging, and also, where necessary, that at the side. The posts are first set up, care being taken to get each post in direct line and to give each post an outward slope at the bottom corresponding to that of the other posts of the sets, the top being spaced by the framing of the timbers, every cap being framed exactly alike. The lining up may be done by sighting or by means of a plumb line suspended from a nail driven at the center of the cap, the cord being made to coincide with a center mark on the graduated staff, as was shown in Fig. 5. If the posts are over 6½ ft. high, it will be necessary to construct a low platform on the floor of the drift to enable the men to work at the cap and bridge.

When the posts and cap are in position the wedge blocks and bridge are placed on top of the cap, the set being made secure by wedges driven at the sides and top. The false set may then be removed by knocking out the foot of the posts, when the lagging will settle firmly on the bridge. If necessary, the wedge blocks in the last set may be knocked out by driving lagging against them, the gap otherwise left being thus filled. Having done this, the sprags are put in place and securely wedged.

When the ground is of such character that the false set is unnecessary, the sprags may be put in place before the main set is firmly wedged, the latter being allowed to "ride" forward just far enough to permit the sprags to clear the shoulders of the daps on posts and cap, when the set is hammered into an exactly upright position. Where the false set is employed it is not always safe to do this, the sprags being secured by wedging. The driving of side lagging is accomplished in exactly the same manner as that overhead, whether the bridge be used in the work or not. It should be understood that the bridge becomes a permanent part of the set and is not removed. The bridge should be not less than 3 in. thick and 6 in. wide, and larger dimensions are frequently advisable.

We have here contemplated the employment of timbers 10 to 14 in. in diameter, which are readily placed by strong, experienced men, the only kind of men, in fact, suited to work in the timber gang. There are mines, however, where the ground is so heavy that timbers 24 to 30 in. in diameter are in use in the main gangways, which it is necessary to keep open constantly. In these sets it is not uncommon to see the posts 8 ft. high and the caps 7 to 9 ft. long at the top, the upper part of the drift at the under side of the cap being 4 to 5 ft. wide inside the timbers. It is manifestly impossible for men to lift from the floor a cap of these dimensions — 8 ft. long and 30 in. diameter, and weighing from 500 to 700 lbs., and, if unseasoned or wet timber, even more — and place it in position on the posts. It is necessary, therefore, to hold the posts in position by spiking braces (lagging) from the last set to the new posts, and also pieces crossing the drift, to keep the posts apart. A platform is then constructed by laying a floor of planks on the braces, which are placed about 30 in. above the floor of the drift (according to the height of the posts), upon which the men may stand while lifting the heavy caps into place and putting the bridge and blocks in position. Of necessity the platform must be well constructed and of sound material, as it must sustain the weight of the cap and that of 4 to 6 men as well. The men of an experienced timber gang work together "like clock-work," systematically, quickly, and with few words, and the heavy timbers are placed, blocked and the job finished in remarkably short time.

The forgoing description is that of driving through easy ground — picking ground, in fact, that not only is readily excavated, but which threatens to drop on short notice, and which does so, unless promptly supported.

The methods here suggested and described will, if properly applied, be found applicable to every such case, but there are times when these simple methods have to be supplemented and elaborated by other arrangements, quite as important to the advancement of the work as the expedients already referred to. We will now consider running ground.

Chapter IV

DRIVING IN RUNNING GROUND

RUNNING ground is the most difficult of all kinds of material through which the miner must at times work his way. Not only does this sort of ground require support after excavation, but it must be held in check constantly while the work is advancing through it. It is sufficiently bad at all times, but when wet the difficulties and dangers are multiplied greatly. Headway through bad running ground can only be made by the employment of close lagging on top and sides. Plank is superior to split lagging in this work, as the plank, being sawed, is smooth and when placed close together leaves no cracks or open spaces through which the fine loose dirt can sift and eventually open a considerable hole, which, weakening the timber support, will finally result in a cave which may be most serious in its consequences. It is imperative, therefore, to timber with extreme care when passing through running ground, and to lag the ground so thoroughly with strong material that there is no likelihood of a run ever starting so long as the timbering is maintained in good condition.

There is a disposition on the part of miners who find themselves obliged to work their way through wet, loose ground, to crowd the work as rapidly as possible and to get through the bad place quickly. It may seem paradoxical to say so, but just the reverse has been determined by experience to be advisable. If the work be advanced more deliberately, it will be found that expense will be reduced, greater security afforded, and in the end better headway will have been made than where the effort to hurry the work has been made.

If the timbering is thoroughly done, and the wet ground is removed by slower degrees, much of the water will drain out of the ground in the vicinity of the face of the cutting, and the ground will in consequence settle more firmly and be more easily removed, with lessened danger of a run.

Breast Boards

In the last chapter the method of passing through loose ground was described and illustrated. To pass through running ground practically the same general scheme is adopted, but in addition to what was described there, the miner employs what are termed "face boards," also called "breast boards." The manner of using breast boards is illustrated in Fig. 11, taken from Bulletin No. 2 of the California State Mining Bureau, a treatise on Mine Timbering, by the writer of these papers. It illustrates a specific instance, which is better than a more general description. The general scheme is that followed by miners almost everywhere, the particular feature displayed in the figure being the use of the "foot blocks" beneath the posts. This was the method of timbering employed by Richard Rowlands, an expert mining engineer of Placerville, California, in driving a drift through heavy running ground in a mine in Sierra County, California. As its employment there was eminently successful, it is advised here as applicable to any situation which involves passing through running ground. As will be observed in the sketch, the posts were not set directly upon the ground, as is usual, but upon wedge-shaped blocks which in turn rested upon double, flat foot blocks, placed, as shown, at right angles to each other. Sills are of no advantage in swelling ground, and of little use in running ground. The foot blocks, however, give the posts a broad and firm support, and as the base is always accessible, any encroachment of the ground from either side or bottom may be promptly removed. The wedge blocks were intended to prevent the posts from taking a greater spread, and in this case experience showed that the idea was correct, as the posts did not shift in the least. In cases where the floor of the drift is firm and does not swell upon exposure, the foot blocks and wedge blocks may usually be dispensed with. It is entirely a matter of judgment.

By carefully studying the sketch, Fig. 11, it will be noticed that the lagging is driven both overhead and at the sides, and that the bridge is employed in both places. Notice also the framing at the inside corner of post and cap — the former is framed with a bevel, the latter at a right angle. The idea was to prevent slipping, but although ingenious, in the presence of heavy pressure it is probable that the advantage of this style of framing, if

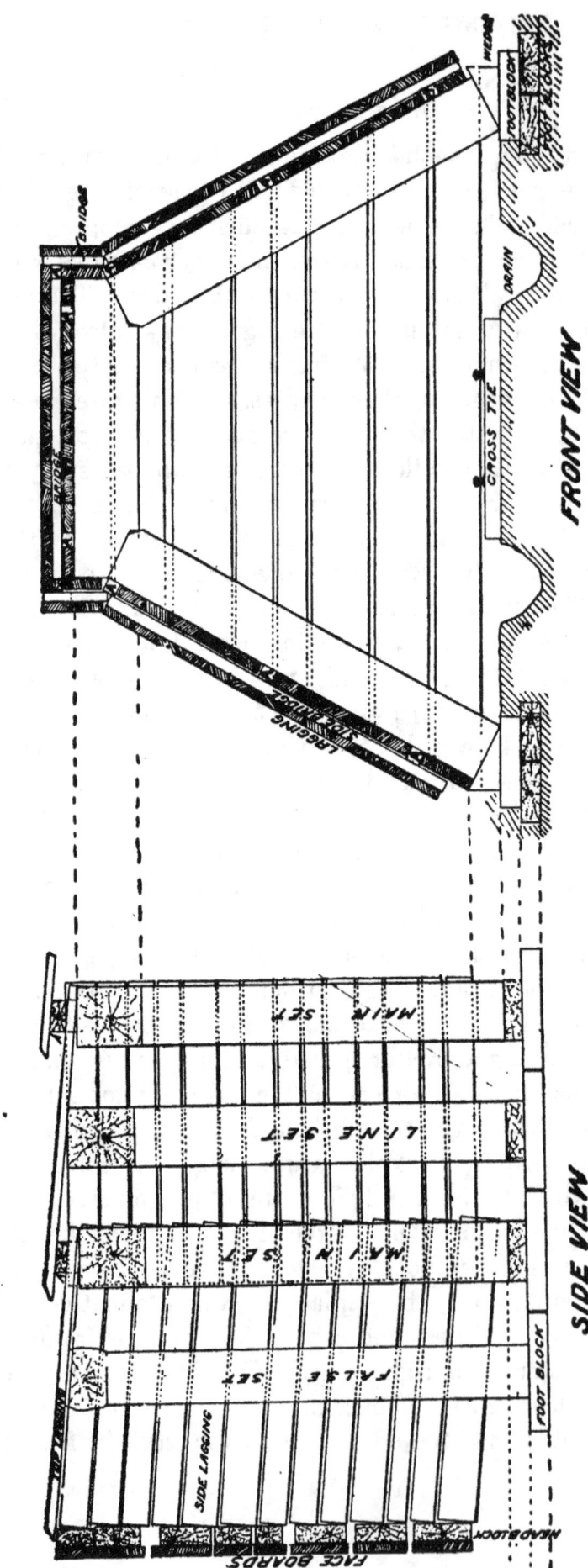

Fig. 11.—Driving with use of Face Boards

it possesses any, would be slight. The system of timbering here illustrated contemplates the use of the false set while driving and of the placing of intermediate sets later, should the weight of the ground on the lagging and main timbers seem to demand it. In this connection it is well to remark that in repairing timbering of any sort in mines, wherever there has been a subsidence and it is required to force the timbers back into place, there is no device that will compare in utility and effectiveness with a hydraulic jack. It occupies small space, is easily carried from place to place, and it has a capacity almost beyond belief.

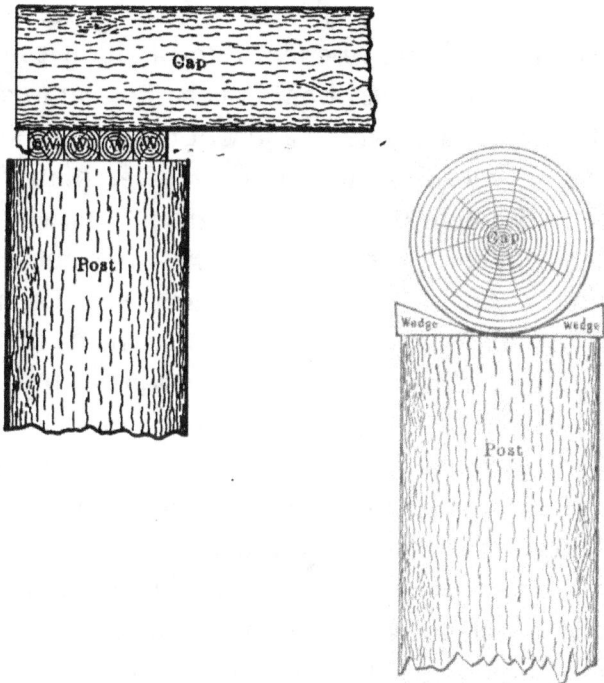

Fig. 12. — The Use of Saddle Wedges

Ordinarily, miners drive sagging caps and other displaced timbers back into position by the employment of "saddle wedges," or wedges driven from opposite directions between the sagging timber and a post or other form of prop placed beneath or opposite it. It is effective, but cannot be compared with the hydraulic jack; nor even with the ordinary jack screw. Fig. 12 illustrates the use of saddle wedges. It should always be remembered that a hydraulic jack is a most useful tool about a mine, and it is surprising how few of these valuable devices are in use for the purposes here suggested.

The employment of "face" or "head boards" is the only device in timbering in mines that makes it possible to pass through running ground. In the instance here illustrated the timbers are placed in exactly the same manner as described in the last chapter, but the face boards and head blocks are an addition to the method there described. Usually, for face boards, 2-in. planks are sufficiently heavy, being generally 6, 8 and 12 in. wide, for convenience in working, the broad plank being placed near the bottom of the working. The head blocks are usually of 4-in. lumber and are 6, 8 and 12 in. wide, to correspond to the width of the face boards. They should be sawed long enough to give them ample bearing at each end. Excavation proceeds by cutting away the ground from beneath and behind the uppermost face board. Sometimes this board can be removed entirely until sufficient ground has been taken out to permit of it being forced forward, but more often it is the better plan to leave the boards in place and gouge out the loose ground as well as possible through the small open space between the uppermost board and that next below. After the upper board has been advanced several inches, keeping the side lagging driven up constantly and held by wedges, the work of excavation will proceed more rapidly. The operation is repeated from board to board, working from the top downward until the bottom is reached. Toward the bottom it is usually permissible to remove the boards, one at a time, as no great amount of ground could run into the face of the drift, but lax methods must be guarded against constantly, and the work conducted, wherever possible, under the direction of an experienced man, or disaster may come quickly and perhaps result in a fatality.

In this manner the work is carried forward, with the aid of the false set previously described, until the full length of the lagging has been reached, when the posts and cap, with the bridges and blocks of the main set, are placed in position. The forward ends of the lagging then rest upon the bridges and are flush with the forward end of the new set. Excavation is then commenced from beneath and behind the upper face board, and the board forced forward as fast as the ground is removed, and the work continued downward, as before, until it becomes necessary to put up the false set. This done, proceed as before, and thus, set by set, the work advances, and if the materials employed are sufficiently strong and properly placed, there should be few stoppages, other than, perhaps, to allow water to drain out of the ground from time to time. When passing through fissures, zones of crushed rock and layers of granulated material, it is well to take precautionary measures, as such ground is always more dangerous than that which is solid and not disturbed by fissuring or brecciation.

Chapter V

STRUCTURAL STEEL IN MINE WORKINGS

Before proceeding further with the methods of timbering drifts, gangways and cross-cuts, it will be well to give some consideration to the increasing use of structural steel in mine workings in this country. Steel has long been employed in underground workings in Europe, where such use has proved the adaptability of this method of support to the conditions there found. The first steel manufacturing company in the United States to make a feature of steel mine supports is the Carnegie Steel Company, of Pittsburg, Pennsylvania. This company has issued a neat little pamphlet on the subject, illustrating and describing "steel timbers" and their use. Two of the most important of the sketches are reproduced here, illustrating the methods of joining the several members of a set. These members are made of a combination of I-beams and double channels, placed back to back. The steel beams here described seem to have thus far given satisfaction, and the probability is that if steel comes into common use in the mines of the West, a system will be evolved which will suit it to all of the varied requirements of mine timbering. The Carnegie Steel Company thus describes its "steel mine timbers" and their employment:

"The use of steel in mines for square timbers and mine props as well is not an experiment. Numerous installations have been made in tubular form in the most important mines of Germany, while ordinary structural material, whether straight or bent, has had large place in the mining operations of England and France. The first use of steel in the mines in the United States seems to have been made by the Susquehanna Coal Company, under the supervision of R. V. Norris, Consulting Engineer, and there is in use to-day in the anthracite region steel timbers which have been in use for twelve to fifteen years in the deep parts of the mines, exposed to constant contact with mine water and without signs of failure or corrosion. Nothing whatever has been done to protect the steel beyond painting it with good heavy coats of paint, and it is safe to estimate that there are now in use three to four miles of gangways timbered with steel. Of course, the use of steel outside the mines in the shape of head frames, breakers, etc., is governed by the

STRUCTURAL STEEL IN MINE WORKINGS

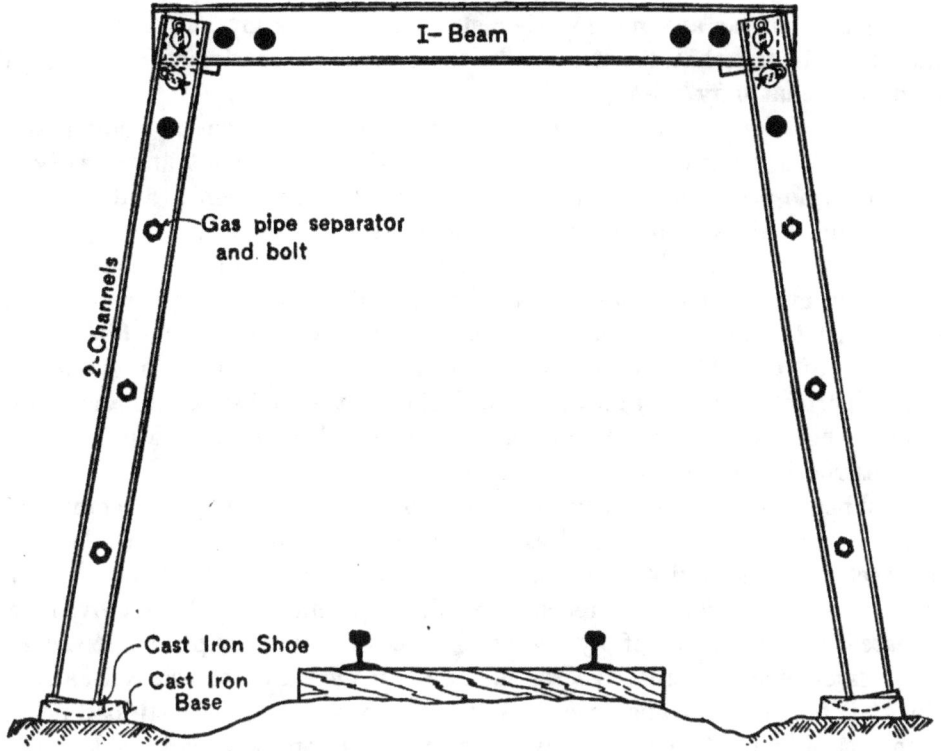

Steel Gangway Support, Style A

Pin Bearing, I Beam Girder, Double Channel Struts, Cast Iron Base

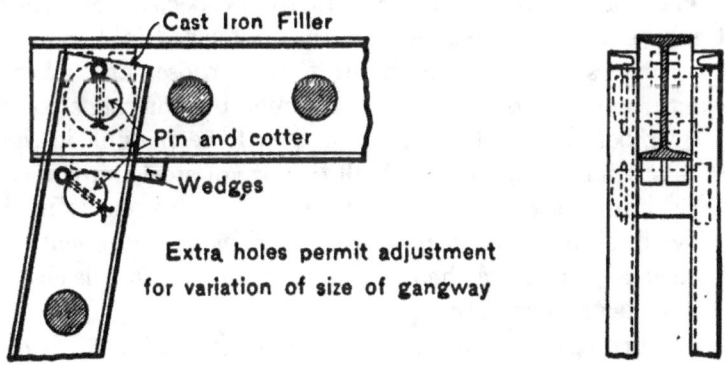

Detail of connection of Girder and Strut

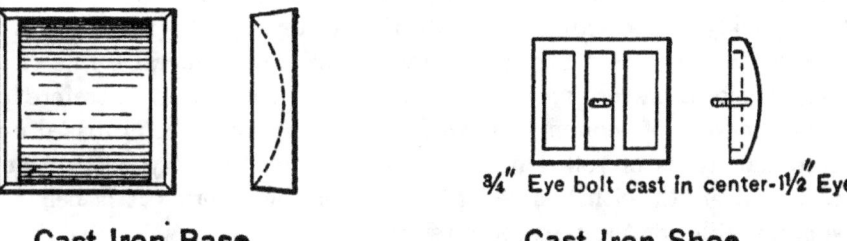

Cast Iron Base   Cast Iron Shoe

Fig. 13

same conditions as are met in other structures above ground, and it is confidently believed that its use inside the mines will satisfactorily meet all conditions that may arise.

"Beyond its long life, steel possesses other advantages as compared to wood. It can be cut to length and fashioned in convenient units ready for erection, and its smaller weight lends itself to convenience and economy in erection. As is well known, enormous destruction of life and property has followed the ignition of dry mine timber by the miner's naked lamp. It is conservatively estimated that the annual mortality in the anthracite region due to fires and accidents is in the neighborhood of one fifth of one per cent of the total number of persons employed. In contrast thereto, steel is perfectly fireproof and its installation at the foot of shafts and slopes, in pump-houses and other points liable to fire, will obviate to a large extent this destruction of life and property.

"When the mine operations in any particular locality are completed, it is not worth the trouble and expense involved to remove the timber that has been installed and it is usually left in the mine at a dead loss. In the forms of construction here recommended, steel can easily be removed and repeatedly re-used, and if by reason of excessive or unexpected stresses it becomes crippled, it still possesses a high salvage value considered as scrap alone. There seems to be no means at hand by which the actual load on the square timber used in the gangways and props can be ascertained, and in the substitution of steel for wood under present conditions, it seems that the only method by which to determine the proper sizes for steel timbers is to proportion them in accordance with the relative theoretical values of the two classes of material. In order to arrive at the proper size of steel timbers to be used, it will be necessary to ascertain from published tables the strength of the wooden timbers ordinarily used for that purpose, and when this is done, to pick out from the tables for steel those members having a corresponding strength. By experiment in this direction, it will be possible in the course of time to arrive at some formulas by which the size of steel timbers necessary for use can be immediately obtained without this comparison; but for the present, it will be more convenient for mine superintendents who are familiar with wood to work from their previous experience in this class of construction and to order accordingly.

"The general arrangement marked, 'Style A' [see Fig. 13] shows that form of construction which was designated by R. V. Norris and which has been used extensively in the anthracite regions, consisting of a single beam collar and double-channel legs, the legs being connected to the beam pins, wedges and cast-iron separators, and resting at the bases on cast-iron pedestals. This form of construction has several features to commend it. It is quickly erected and as quickly taken down. The general arrangements marked 'Style C' [see Fig. 14] represent the same general forms of construction, which however, are simpler of manufacture, and therefore possess a larger measure of economy in the first cost, 'Style C' being based on the use of special forms of rolled material designed particularly with a view to increased strength with less weight. Any of these forms it is believed should be acceptable for the purpose intended."

STRUCTURAL STEEL IN MINE WORKINGS

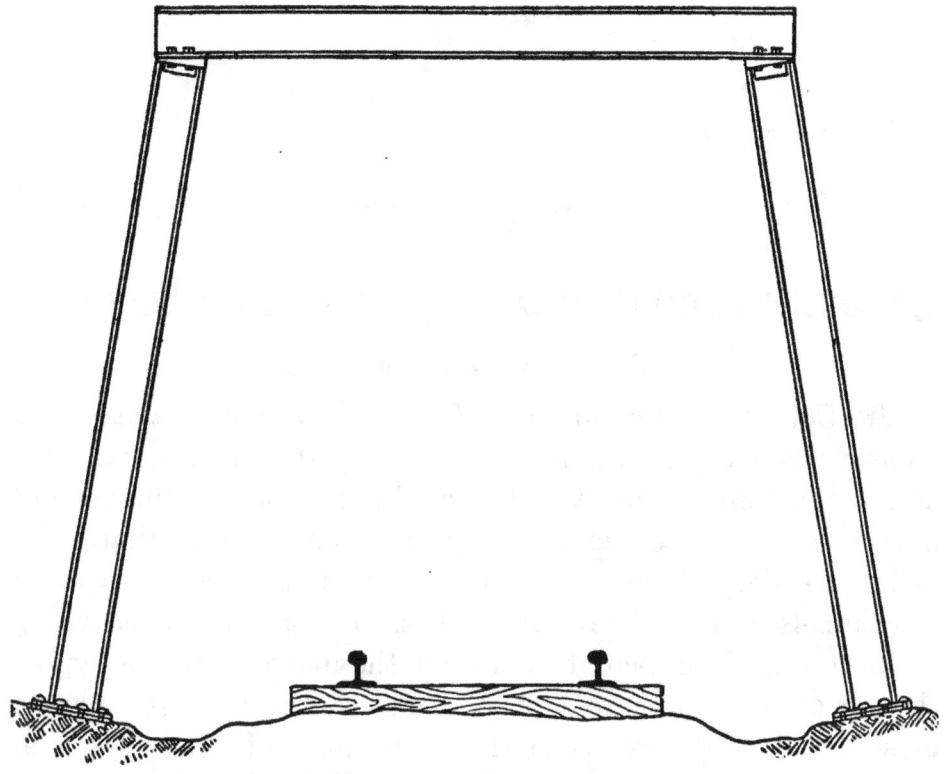

Steel Gangway Support, Style C

H Sections Girder and Struts, Cast Iron Cap, Steel Base

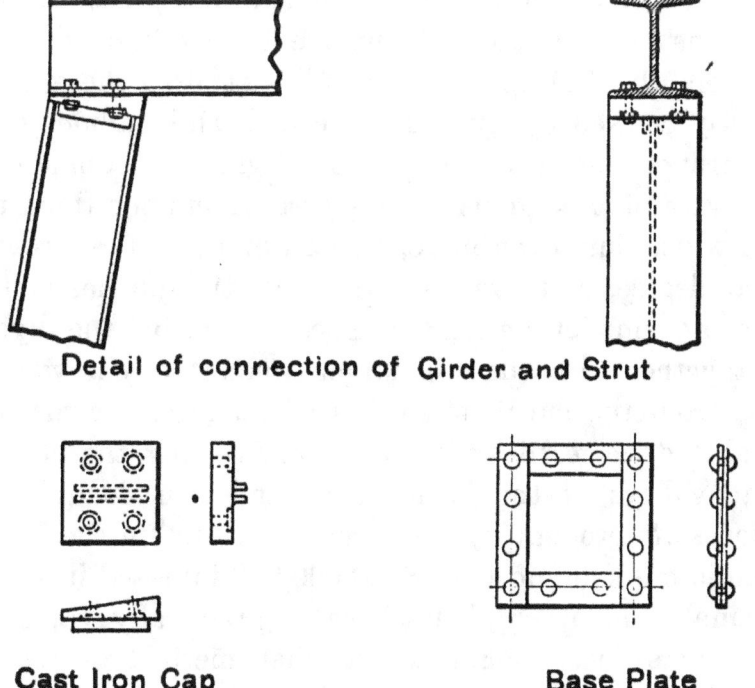

Detail of connection of Girder and Strut

Cast Iron Cap   Base Plate

FIG. 14

Chapter VI

TIMBERING DRIFT GRAVEL MINES IN CALIFORNIA.

Breasting Posts and Caps

In California, the mining of the auriferous gravels of the ancient rivers by what is known as "drifting" has become a large and important industry. These differ from ordinary gulch placers in several respects. They are beds of rivers that existed in late Tertiary times, and flowed as modern streams do to-day, in channels, either in narrow gulches, in broad, timbered valleys, or in courses intermediate between these two extreme types — though of one thing we may be assured, the ancient rivers, for most part, were larger than the rivers in California to-day, and the grades were usually much flatter. These channels are often found firmly cemented and generally covered by a deep deposit of volcanic material, and are often difficult to mine. In the early history of mining in California, these ancient channels were recognized as such, but there seems to have been an impression that they had all formed one great river, named by those early miners the "Deep Blue Lead." At first these channels were mostly worked by hydraulic method, and millions of cubic yards of gravel were torn down by the "giants." Some of these channels were of prodigious size — great rivers one-third to half a mile in width, and even more, though many of them resembled in a great degree, as to width and grade, the gulches of to-day. The vast amount of detritus washed down by the hydraulic miners, together with the still larger amount of silt washed by the rains from the cultivated fields of mountain farms, caused considerable damage to agricultural lands along the streams in the great valleys of the Sacramento and San Joaquin rivers. Injunction suits brought by the farmers resulted in the inhibition of hydraulic mining, and over $100,000,000 invested in hydraulic mines, canals and other plants became practically valueless.

The miners then sought some other method of recovering gold from these ancient rivers. A few mines had been worked by tunneling the gravel on bed rock. This method of mining now was quite generally introduced and the methods improved,

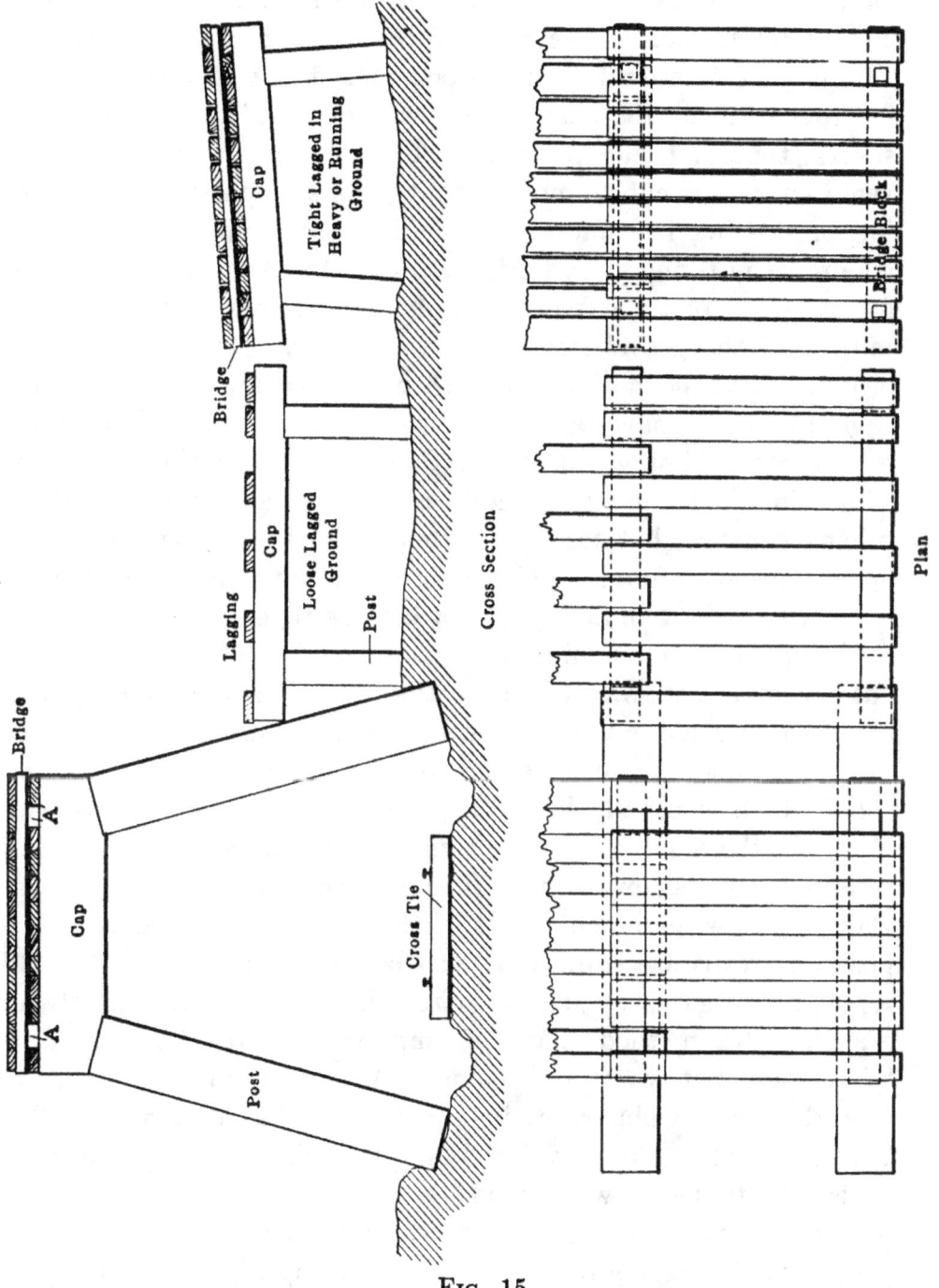

Fig. 15

until drift mining became a large and in many cases a very profitable industry. Often the gravel could be attacked where the end of the channel had been exposed by erosion, or had been cut by the powerful streams from the giants, but in a great many instances the ancient rivers lie buried deeply beneath the lava cap, and often below the drainage of the streams of the vicinity. This latter fact often necessitated the driving of lengthy tunnels through the "rim rock," as the bed rock of the channel is called, in order to reach the channel proper. Some of these tunnels are several thousand feet in length. The mining practice in driving these long cross-cut tunnels is that generally pursued in tunneling or drifting in quartz mines. The same variations are encountered in tunnels run to develop drift mines as in those driven for water or for the exploitation of quartz mines. It is with the mining of the gravel itself that the present chapter deals. In order to make the methods more clear, the accompanying sketches, Figs. 15 and 16, have been introduced, taken from Bulletin No. 2 of the State Mining Bureau of California — "Methods of Mine Timbering." Running ground is occasionally met in drift-mining practice, but far more often it is swelling bed rock that gives the most trouble. This phase of mining has already been dealt with at length, and need not here be repeated.

In opening a drift mine it is customary in the best practice to cut the main gangway (or more than one in a very wide gravel channel) to some extent in the bed rock. This is done for two purposes: first, it aids in equalizing the inequalities of the bed rock, which is usually found to undulate, more or less; second, it makes the loading of cars from the platforms more easy and consequently less expensive. The method of driving these gangways varies with the character of the ground. If necessary, the posts and caps are employed; also lagging, if required. Fig. 16 represents a gangway driven partly in bed rock and partly in gravel. This method may be employed where the ground is fairly good, but where it becomes loose and requires more care, then the method illustrated in Fig. 15 should be employed. The breast — that is, the gravel bed on either side of the gangway — is usually timbered with what the drift miners call "post and breasting caps." These are single posts with a strip of heavy plank or split lagging placed between the top of the post and the roof of gravel or other rock. This is made tight by driving wedges between the cap and the roof. The general manner of breasting gravel is something like the methods employed in coal mines. As the work of breasting proceeds, the cobbles and boulders are stacked behind the miners, these in time forming a broad and firm support to the roof. Often, where no regular sets are required in the gangways, walls are built of the boulders and cobbles lining the gangways. Some mines do not even

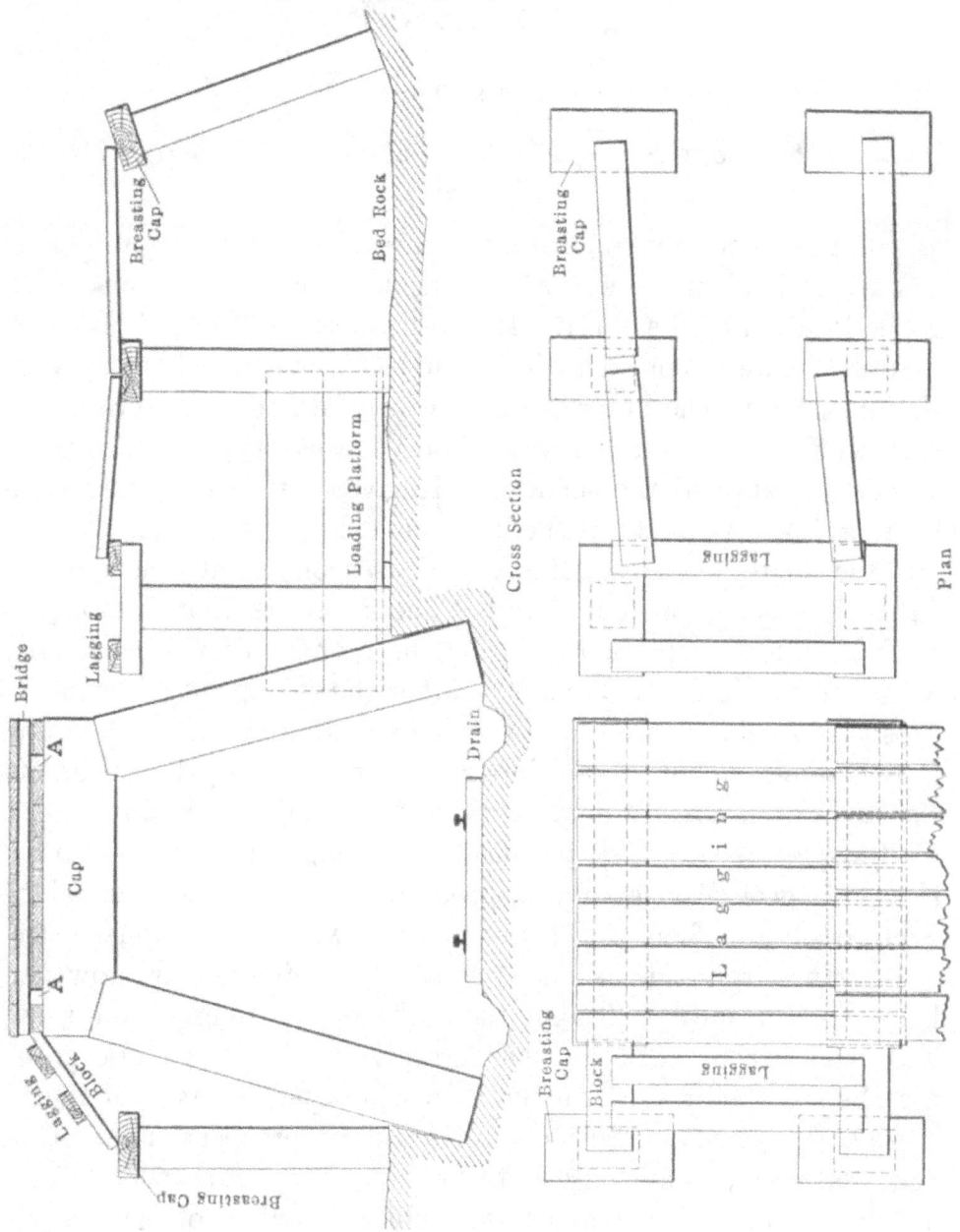

Fig. 16

require the breasting caps and posts, the "gob" or boulders of the fill affording all the support that is necessary. Some ground, on the other hand, requires a comparatively elaborate system of timbering. The two principal methods employed are depicted in Figs. 15 and 16.

Chapter VII

SHAFTS

Location, Kind and Size of Shafts. Substantial Timbering Usually Necessary

Shafts generally require more careful planning than almost any other kind of mine work. The first thing to be considered is the location of the shaft. In most instances the shaft is not a matter of careful consideration with the miner, particularly with the prospector, who begins to sink where he finds good ore at the surface, or at a point where he anticipates striking ore within a short distance of the surface. He gives little thought to engineering features, or to economical considerations, other than to put the shaft down as quickly and as economically as possible with the means at command. The result is that a prospect shaft is seldom useful as a working shaft, though not infrequently a shaft suitable for a working shaft of large capacity is sunk to prospect a vein or ore deposit — a bad practice and one not to be recommended on undeveloped property, unless the probabilities have been already demonstrated in an adjoining mine.

Prospectors are fully justified in sinking small shafts, and it is a good general rule to sink prospect shafts on the vein whenever possible. There are few exceptions where conditions justify a departure from this excellent rule. Not infrequently, however, large and usually well-appointed shafts, thoroughly equipped with hoisting plant and other essentials of a large operation, may be seen being sunk on the merest prospect — little or no ore in sight - to justify more than an ordinary prospect hole. It is needless to say that these enterprises are the outcome of some elaborate promotion scheme, the chief function of which evidently is to enrich the promoters, regardless of the intrinsic value of the property or its prospective worth. Nevertheless, such shafts are often models of engineering excellence.

Prospectors are frequently excellent miners, from the standpoint of economy. They can put down a shaft and support the loose and threatening rocks in a most ingenious manner. Most experienced mining engineers who step upon the platform of a cage or skip over a hole 2000 to 5000 ft. deep without the slightest fear, look down into the dark depths of a prospect hole less than 100 ft. deep with apprehension, and descend in the bucket or upon the rickety ladders with many misgivings. This is only natural. In the former there is every evidence of substantial materials, superior workmanship and engineering skill; in the latter on every hand may be seen the makeshift, the apparently insecure, and an absolute lack of knowledge of anything like engineering practice, though a more careful scrutiny will usually prove that skill has not been lacking. There are no extra or unnecessary timbers. Every loose block is firmly supported, and the ladders, though not models of neatness, are usually strong enough for the purpose for which they are intended. Prospect shafts should be sunk on the vein or in the ore deposit. Good prospects have not infrequently been lost to the original owners by failure to follow the vein. A large amount of work may be done in a shaft off the vein; funds becoming exhausted and no paying ore developed, the miner quits, and another later develops a mine by staying with the ore.

A shaft intended for working purposes can only be intelligently located after the mine has been sufficiently developed to determine the position and extent of the ore. Not only must the position of the ore bodies be known, but their trend as well, and if there are two or more veins in proximity the shaft should be located with due reference to the several veins. The topographical situation is also important. When ore has been hoisted to the surface it must be disposed of — sent to mill or smelter, to railroad cars or to wagons, to be transported to some place more or less distant. Roads must often be considered; consequently the shaft must be started at a point readily accessible by heavily laden teams, both for the quick and economical handling of machinery, timbers and other supplies, and for the hauling away of ore and waste rock. Particularly is it desirable to have the shaft at a place which will afford transportation of ore by gravity to the reduction works. Where possible, there should be sufficient fall between them to admit of the placing of a large ore bin, below which should be situated the rock breaker, discharging into a second bin, from which the ore may be drawn and sent directly to the mill or smelter, if on the property, by gravity. This very important desideratum in the economy of mining is too often overlooked or neglected as of minor importance. If a large tonnage is to be handled, it may mean a vast difference in cost. Assume a mine which will produce 1,000,000 tons of ore. The additional unnecessary expense of five cents per ton in elevating the ore to the mill after it has been raised to the collar of the shaft, when it may have been avoided by locating the shaft at some higher point, means an outlay of $50,000, which would in most cases far more than offset the possible additional cost of sinking the shaft at some other place. The illustration, Fig. 17, shows a shaft located with reference to metallurgical operations.

Another thing to be avoided is the sinking of a shaft in a gulch at a level so low that there is even a slight possibility of a freshet pouring down the shaft, flooding the mine and possibly causing the death of men working below. This caution is the result of knowledge of several shafts so located that this misfortune actually occurred. In some instances shafts are sunk on the side hill sufficiently high to be above high-water mark, but where much surface drainage finds its way through the soil and loose rocks into the shaft. This may be largely, if not wholly eliminated, by building a substantial wall of concrete about the shaft, this wall having a base far enough below the surface to cut off any such superficial drainage. Concrete is also advisable around the collar of a shaft sunk through loose ground where the shaft is not situated in a gulch. There is an increasing tendency to employ concrete in many places underground, its usefulness and economy being constantly demonstrated.

No branch of mining practice has received more careful attention than shaft sinking. It is the most important phase of all mining work. Large working shafts are necessarily expensive, and they should be so made and equipped that they can be kept open as long as there remains ore in the mine that may be extracted at a profit. Engineers have devoted much time to devising methods of shaft sinking, with a view to improving practice while hastening the speed with which the work is accomplished. In many cases the desired result has been attained, but as a rule the expense has not been decreased correspondingly. However, there is one feature in the economy of shaft sinking too

SHAFTS

Fig. 17. — Shaft of the Detroit Copper Co. at Morenci, Arizona. Located with Reference to Metallurgical Operations

often lost sight of, and that is the necessity for as great speed as possible commensurate with good workmanship and safety. Where large capital is invested in a mining operation, the interest charge is an item of no small consequence, and speed in shaft sinking often justifies an increase in expense per foot. This has given rise to the premium system so successfully applied in sinking the deep shafts on the Rand in South Africa. The spur of additional pay for additional work is seldom without the desired effect on the miner, be he American, Cornishman, Mexican, Kaffir, Chinaman, or any other. It is an excellent system when wisely applied.

Prospecting Shafts in Good Ground and Those Requiring Close Timbering. Cribs and Single-Compartment Shafts

Prospectors often sink shafts to surprising depths by means of hand windlass or horse whim, employing little or no timber. Of course this is only possible where the ground stands well. Near the village of Mokelumne Hill, in Calaveras County, California, are shafts sunk in 1851–52, to a depth of 350 ft. or more, which have no timber from top to bottom and never had. The rock was hoisted by means of windlasses, and although these shafts were sunk nearly sixty years ago, they are as good and safe to-day as when they were put down. This is due to the fact that they were sunk in the volcanic tuff that overlies an ancient river channel. These holes are circular, about four feet in diameter, and vertical. In the Calico mining district, San Bernardino County, California, are shafts 200 ft. and more in depth that are without timbers, having been sunk in the early eighties. These shafts are in the rhyolite tuff of that region, and as the climate there is remarkably dry, they are open and in good condition to-day, although they are rectangular or square in section. Doubtless, shafts may be and have been safely sunk under similar conditions in many other places, but ordinarily it is not only wise, but necessary, to use timber in shaft sinking in most rocks (tuffs being about the only exception), and even in this kind of ground it would be unsafe to use such shafts for working purposes without the employment of some timber.

In other places in the desert regions of California, Nevada and Arizona, prospectors have sunk shafts several hundred feet in depth, both vertical and inclined, where only the timber absolutely necessary was put in — each stick being placed to support some particular block or mass of rock. As these several rock masses require support from various directions, the timbers are placed like stulls in a stope and exhibit no regularity of distribution. Often the appearance of these shaft stulls reminds one, looking upon them from above, of a corkscrew. Where the shaft remains constantly dry, and the timbers have been properly placed and well wedged when put in, there is comparatively little danger, but where there are alterations of wet and dry, there arises a great danger, unless the miners frequently inspect the timbering and keep the wedges well driven. The reason for this is, that when wet the wedges swell, and all is secure, but when the atmosphere becomes dry the wedges shrink, and if the necessary attention is not given them the timbers are more than likely to drop out of their own weight, unless the rocks settle more firmly upon them, which immediately creates a new danger in the threatened caving of the ground. All this, however, can be largely, if not wholly, avoided by keeping the wedges tightly driven at all times. Prospectors will generally utilize such timber as is most readily and cheaply obtainable. On the desert, long experience has demonstrated that almost any kind of timber support will do, and consequently in the prospect holes of the great Southwest may be seen old railroad ties, knotty sticks of juniper, scrub pines, scantling, and even yuccas and sage brush — the latter in place of lagging — to hold back small rocks. In the timbered regions, the prospector is more painstaking, for several reasons. In the first place, there are in such regions alternating seasons of wet and dry, and his ground requires more elaborate support; then he has usually an assortment of timber from which to choose, both as to kind and size, but he usually selects that which is most easily handled and most quickly prepared; therefore it is not uncommon to see prospect shafts in such regions cribbed with small timbers or even with poles.

As correct methods of shaft-timbering require considerable framing, which demands suitable tools and no little mechanical skill, we find the prospector "framing" his crib timbers with an axe. The careful and more experienced will strip the bark from the poles or timbers before placing them in the mine; others blaze the side next the shaft; while those who care nothing for the permanency of the work will put the sticks in the shaft without removing any of the bark at all. A crib built with the bark on seldom outlasts two years, and if not renewed the shaft is usually lost.

Near Grizzly Flat, El Dorado County, California, the writer saw a two-compartment shaft nearly 300 ft. deep on the Mount Pleasant mine. Climbing down the vertical ladder of this shaft, it was observed that the timbers, which were 12 × 12 in. and splendidly framed, presented an unusually smooth appearance. At first it was supposed that the timbers had been sawed and then planed, but closer inspection showed that each timber had been hewn by a master hand — probably by a ship carpenter, for no

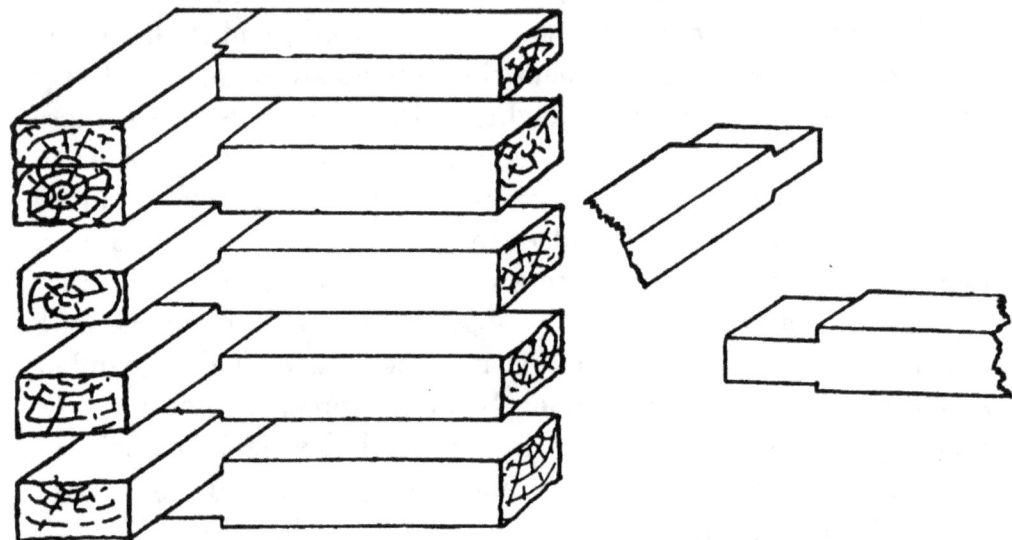

Fig. 18. — Crib of Timbers in Shaft

miner would have ever taken such infinite pains to produce the artistic results.

Cribs are made of either round or sawed timber, both large and small. The round sticks are more commonly seen. As the work progresses, if several feet can be sunk ahead of timbering, a good set of four pieces is placed in position and firmly wedged. Upon these the crib is built upward toward the surface to the collar of the shaft, or to meet the last lot of timbers put in. Excavation then proceeds as before, and when it has gone as far as is considered safe, the crib is again commenced and built up to join the last lot. If the ground is too loose to admit of this, the crib must be built from the top downward, and a set put in and wedged as fast as room can be made for it. Should the ground be so loose or heavy as to render this a difficult or unsafe job, it were better to employ some one of the regular methods of shaft timbering and not to attempt the crib.

The accompanying sketch, Fig. 18, illustrates a crib of sawed timber and the method of framing. It is desirable to leave as little open space as possible back of the crib timbers — or, in fact, of any timbers — in shaft sinking through loose ground. For this purpose rocks, waste timber ends, and often cordwood, are employed. It prevents the ground from "starting," which is very important.

Size and Division of Working Shafts. Drainage

Working shafts for mines are made with one or more compartments. The shaft of three compartments is the most common, though some small mining operations have been successfully carried on through a shaft having but a single compartment. As a matter of course, a large mine, outputting a thousand or more tons of rock daily, might be operated through a single compartment if it were large enough. The only reason this is not done is because of the necessity of sustaining the rock masses through which the shaft passes. Long experience has demonstrated that it is extremely dangerous to leave any large rock surface exposed without providing proper support. If rocks were homogeneous, and not divided as they are into innumerable blocks by cleavage, jointing and other planes, the ground would probably stand very well, and no timber or other support would be required excepting such as might be necessary to carry guides, ladders and pipes; but the fact that all rocks are divided by these planes of jointing, bedding and cleavage, beside more or less numerous fissures, renders it necessary that timber, steel, masonry, concrete, or some other artificial support, be employed to render the shaft safe. In Mexico are some remarkable examples of large shafts. In a few instances these great excavations exceed 2000 ft. in depth and are 20 or more ft. in diameter, being circular or octagonal in form. The rocks are supported by masonry. Hoisting was accomplished by means of horse whims set opposite the several sides of the shafts. These great shafts are among the native engineering wonders of Mexico. Similar shafts have been sunk in the coal fields of northern China. Surely the men who conceived and directed the execution of the work in making these wonderful excavations were engineers of no mean order.

In the United States few shafts have more than four compartments, though there are some in the Lake Superior copper region having six. A surprisingly large tonnage can be hoisted from a depth of 3000 ft. through a well-equipped shaft of three compartments if the proper loading and discharging facilities be provided. In consideration of the fact that nearly 4000 tons have been hoisted daily through a three-compartment shaft from a depth exceeding 1500 ft. for months at a time at Kimberly,

Fig. 19. — A Prospector's Shaft with Horse Whim

South Africa, it would seem that a shaft of this description would answer for most requirements of mines in the United States. It is our opinion, however, that the most complete shaft is one of four compartments. Two compartments for hoisting ore (and waste when necessary); one compartment fitted with a special cage for the accommodation of the superintendent, foremen, shift bosses, pump men, skip tenders, powder men, tool packers and others whose duties require frequent passage up or down the shaft. This cage could also be used for the accommodation of visitors, who are more or less numerous at most mines, and who are also a source of no little expense in various ways. This "service" compartment might be somewhat smaller than the hoisting compartments, and the cage in it should be handled by a separate engine. Through this compartment, also, the pumps could be raised and lowered as required. The fourth compartment should be somewhat larger than either of the others, and in it should be placed the ventilating pipes, water and compressed-air columns, and steam pipes, if steam is employed to operate pumps, which practice in large mines is being generally discontinued, compressed air or electricity having to a great extent replaced steam. If Cornish pumps are in use the pump rod should be placed in this fourth compartment; and here, too, are the ladder ways.

A shaft equipped as here suggested affords the highest degree of utility possible. Hoisting can be carried on uninterruptedly; the executive staff can descend or ascend the shaft at any time without interfering with hoisting operations; the pipes for air and water are in a roomy compartment with frequent platforms, enabling any necessary work there to be done without interruption, which saves time and consequently money. In mines where drainage is accomplished by bailing with buckets or skips, it may be advisable to put the service cage in the ladder-way compartment and place the bailing skip in the third compartment. Where a very large volume of water must be handled by bailing, it has been economically accomplished by providing a shaft of two compartments for this purpose alone. In the Pennsylvania coal region, mines are equipped with such shafts, in which the large bailing skips, operated by electricity, descend, fill and are hoisted and discharged automatically, no attendant being required, the engineer on duty attending to the entire electric installation connected with the mine, having no special work to perform, so far as the bailing skips are concerned. These skips are run in balance, and operated night and day regularly, hoisting and discharging their load of water. A skip of 1500 gal. capacity, run at a speed of even 1000 ft. per minute, will raise a very large amount of water fron a mine 3000 ft. deep in 24 hours. Generally speaking, bailing is a less expensive method of draining a mine through a shaft than by any method of pumping, if the bailing installation be properly equipped. An inefficient bailing plant, however, will prove an expensive method of handling water.

Positions of Temporary Hoisting Plant and Permanent Plant

A word about the sinking plant and we will then consider the various means of shaft sinking, the methods of doing the work, and of timbering under the conditions usually met in Western metal mines. The sinking plant should be of such capacity as will permit the shaft to be carried to a reasonable depth, say 400 to 500 ft. This machine, of about 10 to 12 h.p., must have double cylinders, and should be in every way a first-class hoisting machine, and one which can be absolutely depended upon to work promptly and safely. It is a mistake to think that any kind of an engine will do for shaft sinking. There is no more dangerous branch of mining work than that of shaft sinking and the men below should be safeguarded as far as possible. When a depth of 400 to 500 ft. has been attained, it is good practice to replace the small hoist with one capable of going to 1500 or 2000 ft., the small machine being placed under cover to be used later underground in sinking winzes, or to deepen the shaft below an established level. The machine taking its place should be of 25 or 30 h.p., not necessarily a fast-winding engine, but a capable and safe one. With this the sinking may be continued to 1500 ft. or more, levels opened, and connections made. In the meantime, the knowledge gained of the actual resources and probable life of the mine enables the management or consulting engineer to determine the character of the permanent plant and provide for its installation. In anticipation of this, it is necessary at the time of commencing the shaft to select a site for the temporary hoist which will later admit of the large, permanent hoisting machinery and head-frame being placed in position without interfering with the operation of the smaller plant, which may be run almost up to the moment of placing the new plant in operation. With this in view, it is good practice to place the temporary hoists — both the first and second plants — either on the opposite side of the shaft from the site of the permanent plant, or at a point opposite the end of the shaft. In some cases the temporary plant has been set between the shaft and the site of the permanent plant, the smaller engine being set opposite the compartment in which the service cage is afterward placed, this engine, or another, being used for operating the service cage. As it is never called

SHAFTS 53

upon to hoist a very heavy load, this engine would probably have sufficient power to operate the service cage to a far greater depth than would be economical in hoisting waste or ore.

Fig. 20. — Steel Head-frame at Old Dominion Mine, Globe, Arizona

The topography about the shaft will, of course, be a factor in determining the position of the hoisting engines, and the height and shape of the head-frame, but the importance of future permanent installation cannot be neglected, and it is the part of wisdom to plan from the beginning with these facts in mind. Fig. 20 illustrates the modern steel head-frame at the Old Dominion Mine, Arizona.

Size of Shaft Compartments

Not less important than the number of compartments in a shaft through which a large amount of rock is to be hoisted, is the size of the compartments. In California practice it has been the custom to make shafts $4 \times 4\frac{1}{2}$ or 4×5 ft. clear, inside the timbers, the greater number of shafts having two compartments. This is antiquated practice and has in more recent years been succeeded by shafts of somewhat more liberal dimensions, usually with three compartments. It should be remembered, however, that there is probably not a mine in California where 1000 tons of ore are hoisted through a single shaft in 24 hours, whereas there are mines elsewhere where two to four times this amount of ore is hoisted every day through a single shaft. This is the case at Butte, Montana, at some shafts in the Lake Superior Copper region, the gold mines of the Rand, South Africa, at the Kimberley diamond mines, and at a great many coal mines in Eastern states. It is therefore no novelty to hoist 2000 or 4000 tons of rock or coal through a single shaft in 24 hours, but this feat demands a shaft with at least two hoisting compartments of liberal size.

On the Rand, some of the earlier large shafts were made from 4×6 to $4\frac{1}{2} \times 6$ ft., according to Thomas H. Leggett, but these dimensions were found inadequate, and the later shafts were 5×6 ft. with a pump and ladder-way compartment $6 \times 6\frac{1}{2}$ ft. For large tonnage, not only large shafts, but fast-winding engines, are essential. Engines are in use in scores of places that wind a load of from three to ten tons at speeds varying between 1000 and 6000 ft. per minute. In California, the custom of sinking working shafts of relatively small sectional area is no doubt due to the fact that many of the most noted shafts in that state are on the Mother Lode, and the heavy, swelling ground in those mines has led the mine managers to sink small shafts to enable them to better hold the ground. Most of the old shafts were sunk in the fissure, and few miners, who have never had experience in these shafts, realize the difficulties to be overcome, not so much in sinking them originally as in keeping them open later. Among the noted old shafts of the description mentioned as on the Mother Lode of California are the original Keystone, the Wildman, the Eureka, the Central Eureka, the south shaft of the Kennedy, the Hardenburgh, the original shaft of the Gwin mine, and several others of less note. All of these shafts were sunk in the fissure, and each of them proved a source of constant expense in keeping them open.

In more recent years it has been considered the best mining practice to sink working shafts in the Mother Lode region of California either vertically, starting in the hanging wall country, like the Kennedy east shaft, the Gwin mine east shaft, the Oneida east shaft, and the Cross shaft of the Utica-Stickle mine; or inclined, in the country rock, as in the case at the Argonaut mine near Jackson. This shaft, commencing in the hanging wall, is sunk at a uniform angle of 63 degrees, in which it has a decided advantage over those sunk in the fissure, which always varies more or less in the angle of dip, resulting in an unsatisfactory shaft. It is advisable, therefore, in view of what has been learned from the practical and expensive experience of many careful operators in widely separated parts of the world, to make working shafts of liberal size, and to avoid sinking them in a vein or fissure if future trouble is to be avoided. The first cost of sinking a shaft through country rock is almost always more than where the sinking is done in the vein, but if this fissure with depth should prove to be in swelling ground, the extraordinary expense of keeping the shaft in repair within the first five years would more than meet the increased expense of sinking in the hard country rock. The methods adopted to keep open the shafts sunk in heavy ground will be taken up later and explained in detail.

The Collar

When the site for a shaft has been selected with due regard to the various considerations already set forth, the first thing to be done is to take the necessary steps to make the collar of the shaft permanently secure. It is not always possible to locate the shaft at a point where solid rock outcrops. Sometimes loam, clay or gravel is found to a depth of several feet where it is desired to start the shaft. The writer sunk a prospecting shaft of two compartments in Tuolumne County, California, several years ago. This shaft was started on a flat, where the surface material, to a depth of 13 ft., was loam and clay. This earth, when dry, was hard enough to make it economical to blast it with low powder, but when wet a ton might easily ooze through a knothole. Necessarily, the collar of this shaft needed careful preparation, and steps were taken to render the subsequent work below secure. Heavy mudsills were laid on a bed of coarse broken rock, and upon these sills the main supporting timbers of the first few sets were placed. The shaft was carried to a

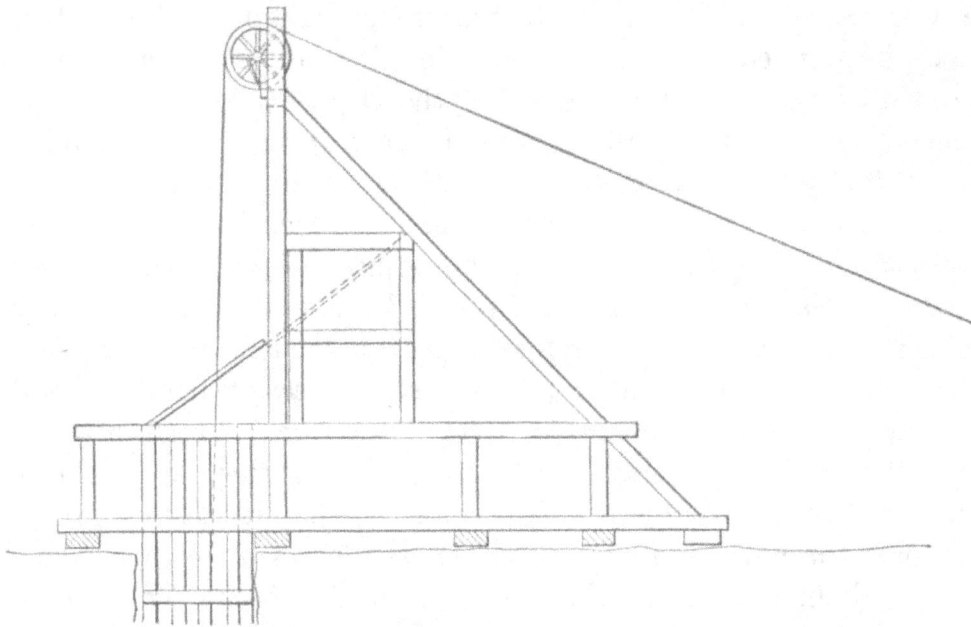

Fig. 21. — Head-frame built on Framework to secure Dump.

depth of 20 ft. and the sets put in place, the wall plates being suspended by permanent straight hanging-bolts made of 1-in. iron. When in place, the space between the lagging (of 2-in. plank) and the earth and rock walls was filled in with cordwood and broken rock, but I would never advise a repetition of this, nor repeat it myself. In that instance it was done to save the shaft, as a heavy and unexpected rain threatened to destroy all that had been done by the caving of the soft, wet clay. Prompt filling saved the shaft, and there was little choice of method. This was a prospecting shaft, and all was done there that the situation seemed to demand. As the surface about the shaft was nearly flat, there was no suitable place to dump rock; consequently a timber frame was built, raising the collar of the shaft about nine feet above the surface, and around this the rock from the shaft was dumped, the head-frame being built upon the timber frame. (See Fig. 21.)

In the case of a large and expensive shaft, planned to go to the depth of several hundred or several thousand feet, nothing would under any circumstances justify any such makeshift method as that above described. It may occasionally be found necessary to carry the shaft up into the head-frame for the purpose of securing dump room, or for some other purpose, but the collar of the shaft must be made permanently secure. The best manner of doing this is to excavate a pit to bed rock in the soil, clay, or loose rock, of such size as will admit of the building of a solid wall of concrete or masonry about the proposed shaft. The inside dimensions of the structure thus built should be such that the regular shaft sets may be placed therein, in perfect alignment with those that are to be placed in the shaft beneath. This concrete or masonry wall should be set deep enough in the solid rock to act also as a dam to the passage of surface water, thus keeping the shaft from being flooded from this cause during the wet season. The wall also affords a substantial foundation for the head-frame — the walls being built back from those surrounding the shaft, where needed, to support the superstructure — head-frame, ore bins and often the rock breaker as well.

When the wall has been completed, the first shaft timbers are placed, extending beyond the limits of the shaft, the ends resting solidly upon the wall. Either the ends or the wall plates may be so disposed. In some cases both ends and wall plates are thus extended, being held in position by drift bolts. Care must be taken in the planning of this structure that the uppermost timbers of the set are not higher than the level of the collar of the shaft. If both wall and end plates are extended to rest upon the wall, it is the better plan to place the ends underneath, the wall plates resting on top of them; then a "filling piece" may be let in on the top of the ends, between the wall plates, thus bringing all timbers flush at the collar. Of course, the timbers may be framed so as to bring them flush, but it seems the best practice to allow to these surface timbers their full strength.

Hanging-bolts

Having placed the collar set, the sets beneath may be suspended in position by means of straight hanging-bolts, which are to remain permanently. One important matter in the use of hanging-bolts, either the permanent straight bolts or the sets of hooks, is the use of washers of large diameter and thickness. Never use a washer less than four inches diameter, even where small timbers are used — 8 × 8 in. for instance, and if the timbers are larger use 4½ or 5-in. washers. Often heavy downward pressure develops after a shaft has been sunk some time, and the stress upon the hanging-bolts causes the washers to sink into the pine wood from ⅛ to ½ in., so that a shaft may settle at one end or side, or all round, for that matter, an inch or two therefrom in four or five sets. When this occurs the timbers may sometimes be forced back into position by use of the hydraulic jack, or even the ordinary jack-screw, but it involves expense and loss of time, which may have been avoided by taking the precaution to use large washers. This suggestion is the outcome of experience in the use of small washers.

Another precaution is to use hanging-bolts of not less than three-quarter iron — seven-eighths being better; and where the shaft is large and the ground believed to be heavy, use 1-in. iron. This admits of the cutting of heavy threads which will not readily "strip," and the hooks will be far less likely to straighten out or break, as they sometimes do when light iron is used for bolts.

The Cross-Head in Sinking Shafts with a Bucket

Before proceeding with the work of shaft excavation we will first consider some of the methods of performing this work in an expeditious and economical manner with reference to surface equipment. Hoisting from vertical shafts, while sinking is in progress, is generally done with a bucket, though occasionally with a low skip. The skip is more commonly used in inclined shafts than in those that are vertical. The bucket has advantages over the skip, however. It is lighter, and therefore easier to handle. It may be placed anywhere on the bottom of the shaft while being filled by the shovelers, which is not the case with a skip. Where the latter is used in a three-compartment

SHAFTS

shaft, the skip should be operated in the center compartment. In shafts where cages and skips are used, hoisting of rock from sinking is often done by a bucket swung beneath the cage or skip. Of course, this necessitates sufficient clearance room in the head-frame. As buckets sometimes are carelessly started from the bottom, they swing about in the shaft and become a great menace to men below. The danger from this source may be obviated in a great degree, if not altogether, by placing guides in the shaft and employing what is known as a cross-head, which, running in guides, prevents the bucket from swinging dangerously in the shaft.

The cross-head is a simple frame of timbers which may be easily made by any carpenter, or even a miner handy with tools. The ironwork can be turned out by any mine blacksmith. They are made after various ideas, almost as various as the number of men who make them. One point, however, must not be forgotten. The cross-head should be twice as high as it is wide, or it may jam against the guides and cause a serious accident; such things are not unknown. Some take two pieces of 4×6 in. lumber, in length a little short of the distance between the inner faces of the guides. These are placed, say, 6 ft. apart. To each side of these are spiked or bolted cross-pieces in the form of the letter X, the ends extending beyond the face of the guides, forming a groove which keeps the cross-head in place. This method of building a cross-head is not as strong as some others.

In the opinion of the writer the most simple and enduring cross-head is made by forming a rectangular frame of four pieces of lumber, carefully sawed and strengthened by iron bolts passing vertically through the structure from top to bottom, the horizontal pieces, top and bottom, being 4×6 in., the wider dimension being uppermost — the size of these pieces being determined by that of the guides. Between these insert two 4×6 in. uprights. A small mortise may be made in the horizontal members, into which may be fitted the tenons framed on the uprights. Auger holes are bored at the center of each horizontal stick, through which the rope is passed. The grooves may be made by securely bolting to the four corners on each side $\frac{1}{4}$-in. steel plates which project at least 4 in. at the side of the guide timbers. Or a piece of channel steel may be firmly bolted to both edges of the cross-head. These bolts require double nuts,

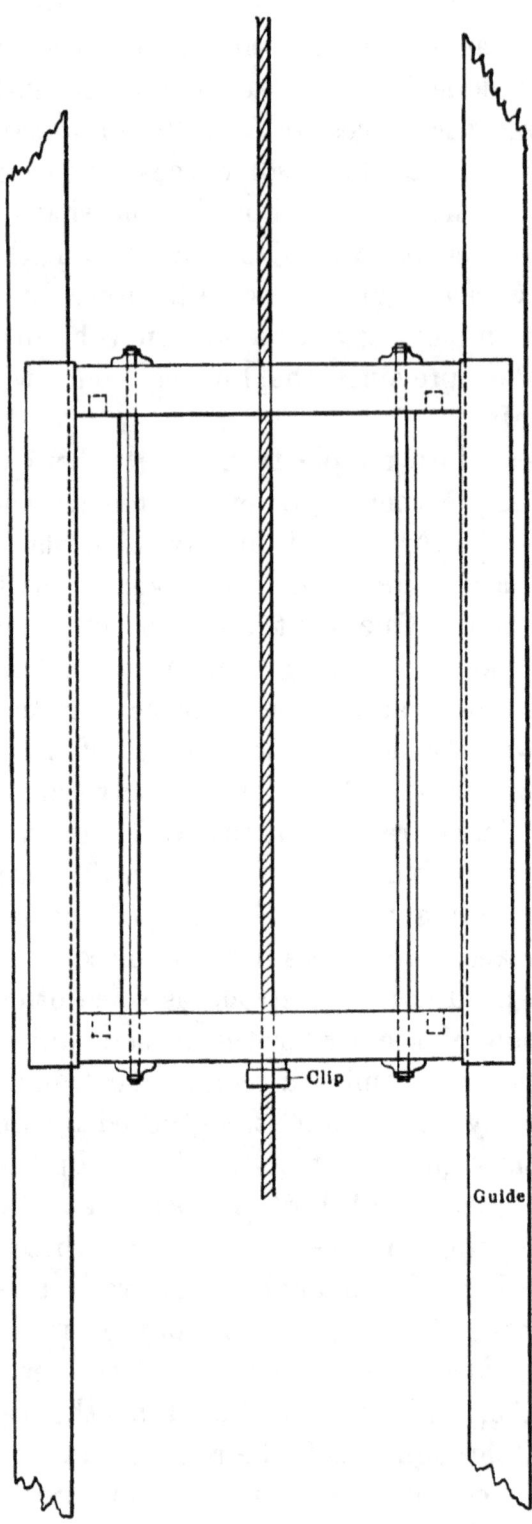

Fig. 22

SHAFTS

so that they may not become loosened by use. The upper and lower edges of this guide runner should be turned back from the guides so as to avoid any inequalities on the guides. The accompanying sketch, Fig. 22, shows the method of constructing a cross-head of this kind. Underneath the lower bar will be observed a clip upon the hoisting rope. This should be securely fastened. Its function is to lift the cross-head when the bucket is hoisted. The clip should be placed so far above the bucket that it will be out of reach of men riding the bucket. Failure to anticipate this has caused injury to men by their being struck by the cross-head.

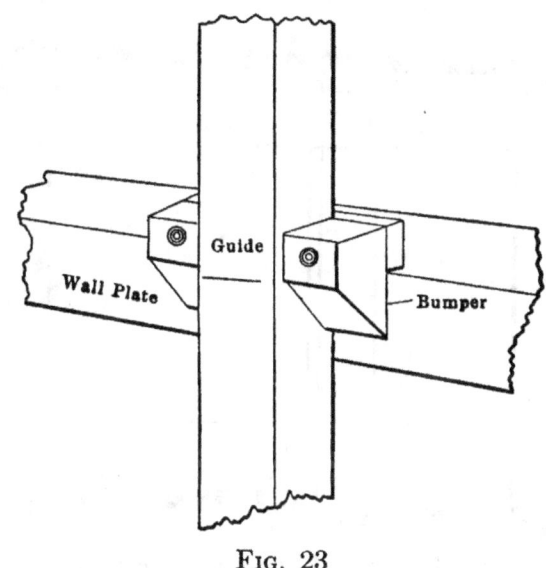

Fig. 23

To prevent the cross-head from dropping off the guides at the bottom of the shaft, blocks should be spiked to the shaft timbers on each side of the guides, and upon these the descending cross-head will rest when the last set of timbers is reached. An improvement on this suggestion would be the construction of a movable stop or bumper, by securely bolting two blocks to a piece of 3 × 8 in. lumber, in such a manner that the bumpers may be placed like a saddle around the guides, the 3 × 8-in. piece resting on the end plate or divider of the shaft set, as shown in Fig. 23. As the work of sinking progresses, these bumpers may be moved downward, set by set, as often as necessary.

Stull Methods in Many Small Mines of the Cobalt District, Ontario

Mr. Algernon Del Mar has contributed the following regarding the methods of timbering small shafts in hard rock in the Cobalt region of Ontario, Canada:

"Shafts for prospecting differ in size and equipment according to the financial abilities of the operators and the size and character of the veins and the walls enclosing the veins. Many of the shafts in the West are 4 × 4 ft. inside, and I had occasion to visit one in Norway 20 × 20 ft. inside. These are extremes, and between these there is a safe mean. If the property under

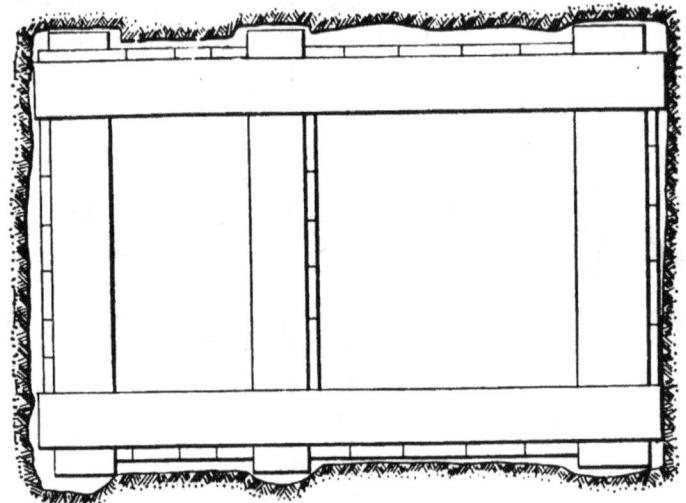

Fig. 24. — Small Shaft in hard Ground

consideration has a showing warranting development to a moderate depth, the shaft should be timbered, or the ground taken out to allow timbering when it is necessary, unless the ground is exceptionally firm, when end stulls for guide timbers, or ties for track, will alone be necessary. A 4 × 5 ft. shaft will do very well for 50 or 100 ft., and when timbered will give a space 3 × 4 ft. for a bucket, but no space for a ladder-way. A 6 × 8 ft. shaft, while wide enough, is a little short in length for a good second compartment. A better size is 6 × 9 ft. outside. This will permit, with 10-in. timbers and lagging, a compartment 4 × 4 ft. for bucket, skip or cage, and, with 12 in. for brattice between compartments, will give one compartment 2 × 4 ft. for ladder and pipe-way. A two-compartment permanent shaft should be a little larger, but a 6 × 9 ft. shaft will answer for prospecting to 400 or 500 ft., and if properly equipped with hoisting machinery should supply a 40-stamp mill.

"The timbering for this shaft will vary with the different kinds of ground encountered, and also with the inclination of the shaft. If the ground has a tendency to cave or run, it will have to be cribbed; if it stands fairly well, it may be timbered with square-sets lagged on the sides that are loose. If the walls cement on exposure to air, as in the desert regions of the West, very little timbering other than the collar sets is required. Where the two walls are firm and not easily broken and the shaft vertical, I have found a compromise method to answer the purpose admirably, and with less expense than the ordinary square-sets. This is to

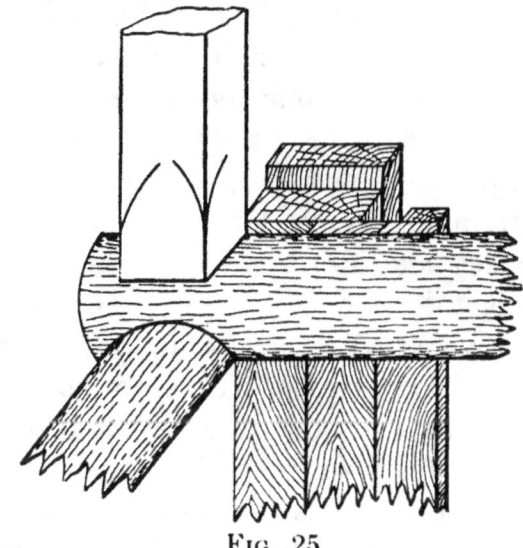

Fig. 25

cut hitches for the stull timbers and on them cut a dap in which rests the wall plate. Place corner posts, wedge and lag in the usual way. This sort of shaft is preferable to the ordinary square-set style when sinking in ore, for the ground may then be stoped around the shaft without displacing the timbers, for the stulls are firmly wedged against the walls. I found this method to be quicker and requiring less knowledge on the part of the miners than placing square-sets, for the hitches could be cut while the machines were at work on the other side of the shaft, and the timbers could be wedged in place without interrupting the work at all. The sketches, Figs. 24 and 25, show the practice in Cobalt district, Ontario, Canada, where the wall rocks are hard. The method has been found satisfactory in that district."

Chapter VIII

BUCKET DUMPING

Methods at Vertical Shafts

HAVING described the cross-head, its usefulness and the manner of constructing it, we may now consider the various means employed to dump buckets upon their reaching the surface. At many mines the "trip-rope" is employed, and this device is certainly a convenient one, as it can be quickly and safely attached and the buckets promptly dumped. All rock buckets have, or should have, a ring securely attached to an eye in the bottom. The trip-rope is suspended at some point overhead — this point is determined by the direction in which it is desired to dump the bucket, and also by the distance from the shaft at which the rock is to fall. It may be dumped directly into a car, into a bin provided with a chute, or on the ground, as may be desired. Often it is necessary to hoist both ore and waste from the shaft, and as a matter of course it is desirable to keep ore and waste separate, when possible. To do this it is not an uncommon practice to have two trip-ropes — one in front, the other in the rear of the shaft, the ore being dumped at one place, the waste at the other. Where the rock is dumped into a bin, it is generally necessary to dump both ore and waste into it, so that whichever is sent up first must be removed before the other class of material is hoisted. This can usually be arranged without difficulty.

The employment of the trip-rope calls for an attendant to be always on hand to close the door over the shaft (it is necessary that every vertical shaft or those sunk at a high angle — over 40° — be provided with a door to insure the safety of men working below), to put the hook at the end of the rope into the ring beneath the bucket, and to raise the door before the bucket can again descend into the shaft. At many shafts this is not an undesirable feature, as the top-man is at other times otherwise

BUCKET DUMPING

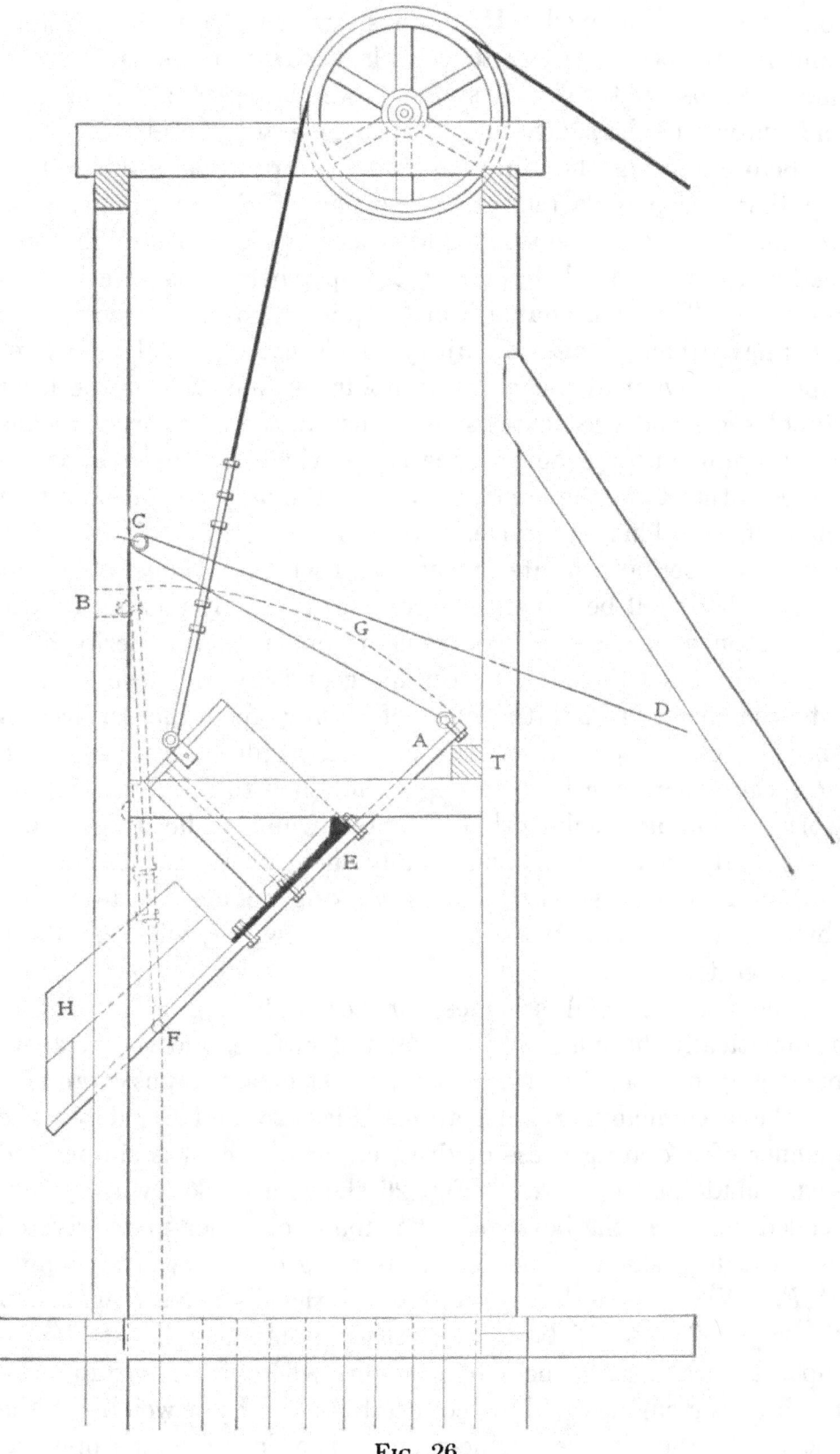

Fig. 26

continuously employed. He assists the carpenter in framing timbers; or acts as helper to the blacksmith; trams cars to the dump; keeps up the fire if steam power is used; brings lagging and timbers to the shaft, and is made generally useful.

Before leaving the top-man and the trip-rope it is well to say that the hook attached to the end of the rope should be provided with an extension handle, as it were — a straight projection from the hook beyond the loop where it is attached to the rope. This will enable him to quickly catch the hook into the ring without danger of injury to his hand. While the trip-rope is a convenient means of dumping buckets, and is the least troublesome and least expensive to install, there are other means of dumping buckets automatically in which no top-man is required, other than the engineer, who is the only person about on the surface. This, of course, applies to prospecting shafts being sunk with, perhaps, limited means, and where it is desirable that every dollar shall be expended economically. By putting a little more money in the top arrangements one man can perform all the work of hoisting, bucket dumping and lowering the buckets into the mine without the need of calling upon the services of any one else employed on the surface. Indeed, we have seen prospects where one man on top found time to do all of the top work, including blacksmithing, timber framing, hoisting, tramming, and general work, but usually such a man has a financial interest in the enterprise, and is working no more than two or three men on a shift below. It shows, however, what a willing man can do.

There are several schemes for accomplishing this work of automatically dumping a bucket at the surface, and the arrangement may be placed in either a four-post or a two-post frame.

The accompanying illustrations, Figs. 26 and 27, depict the manner of operating these devices, either of which is simple and easily made and operated. Fig. 26 shows a bucket with an extended chime at the bottom. It is made of boiler plate, riveted in the usual fashion. The door A is hung on heavy strap hinges at F. When open the upper edge of the door rests against a timber at B. It will be noticed that to this door is attached a rope, D, leading to the engine-room, where it is within easy reach of the engineer. It is counterbalanced by a weight at the end of the rope in the engine-room. The door remains open at

BUCKET DUMPING

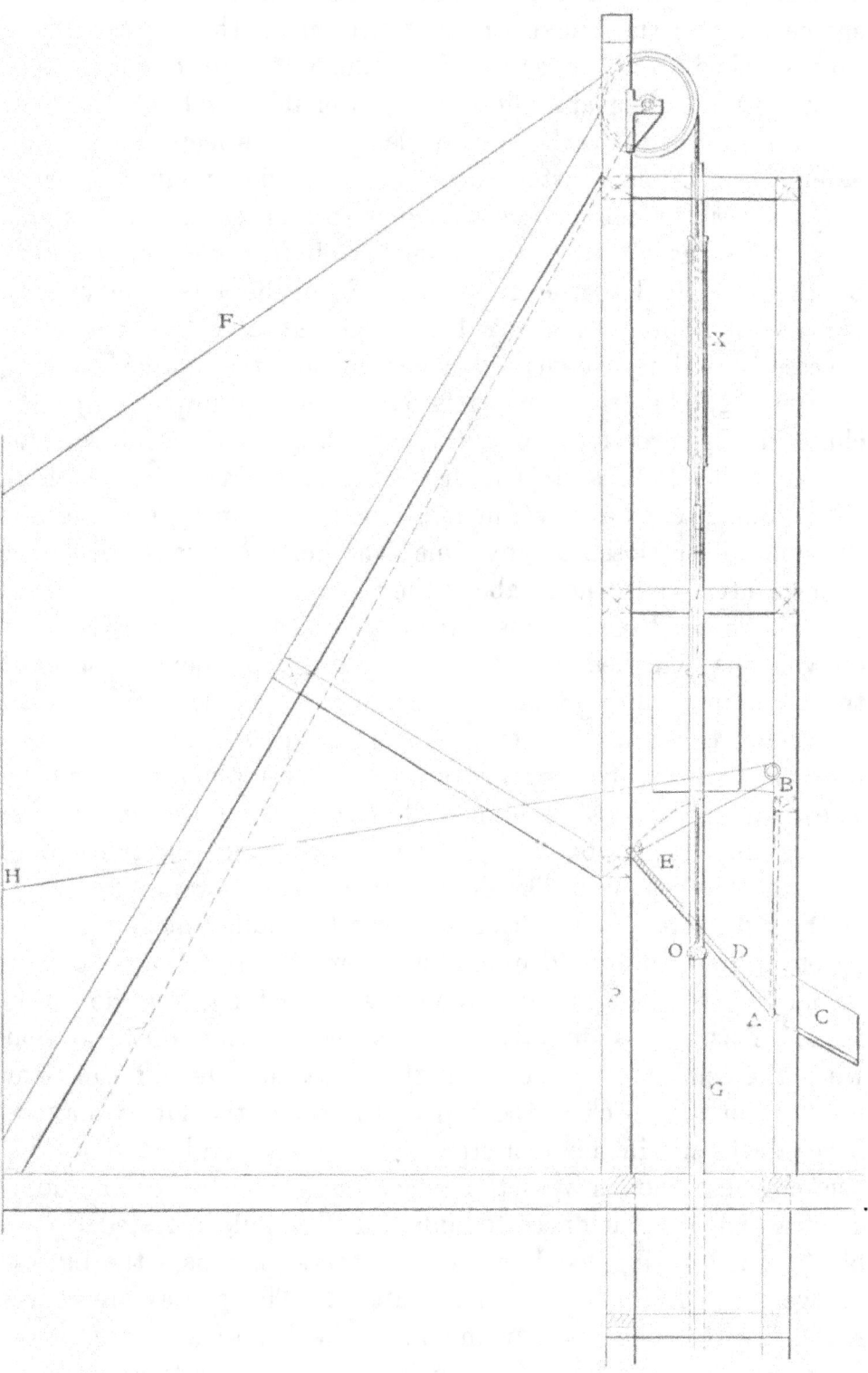

Fig. 27

all times, except when a bucket is about to be dumped. The engineer raises the bucket until it will clear the door. Attention is called to the necessity for sufficient head room between the top of the door and the sheave, to enable the bucket to clear the door. When the bucket is clear of the sweep of the door, as indicated by the dotted curve G, the engineer gives the rope a quick jerk, which causes the door to start forward. Passing its center of gravity it drops across the shaft, resting on the timber at T. Bolted to the upper side of the door is a heavy iron strap with a projecting lug 1 in. high, at E. The bucket is lowered and slides downward a few inches, the chime catching on the lug, and the bucket is overturned, dumping into the chute H, or directly into a car standing at the chute. This device works well, and is in use at many places, but I think the dumping scheme shown in Fig. 27 is better, in that it permits the door to be closed at any time, whether the bucket be below in the shaft, or suspended above the door.

This figure, No. 27, illustrates a two-post frame, to the front of which is attached a structure of lighter timbers to support the door and chute. The cross-head is at X, with the bucket suspended beneath. Under the bucket, and secured by a short rope or chain to the ring in its bottom, is a block of wood, O. The apron or door D is hinged at A, resting at the upper edge on the timber just back of E, when down, and on the timber beneath B, when up. The door is raised and lowered by means of the light rope H, which passes over the pulley at B. In this apron at its center, and extending from its upper edge to near its middle, is a slot, indicated in the shaded part of the apron at E. This slot is directly opposite the hoisting rope, so that when the bucket is raised from the shaft to a height sufficient for the bottom to clear the top of the door, the latter may be lowered, the slot in the door straddling the rope under the bucket. The engineer then lowers the bucket until it reaches the door, down which it commences to slide, but is finally arrested by the block, which, being too large to pass the slot, causes the bucket to overturn, dumping into the chute C. The guides are represented by G. The main members of the head-frame are represented by P. The line of the hoisting rope from the sheave to the winding reel is seen at F, and R is the line of resultant strain. It is suggested that the door, as shown in the drawing, is inclined at too low an angle. Not that it will fail to cause the bucket to dump, but for the reason that it requires the rope beneath the bucket to be unnecessarily long, so that the bucket in dumping drops too far down into the chute. If the door were 2 ft. higher, being still hinged at A, it would undoubtedly work better.

When the bucket has been dumped, the engineer raises it until its bottom is above the top of the door, when the latter may be lifted and the bucket again lowered into the mine.

Methods at Inclined Shafts

We have described at some length the use of the cross-head and devices for dumping buckets at vertical shafts. Probably by far the greater number of shafts — prospecting shafts, at any rate — are inclined, and automatic dumping devices are as necessary at inclined as at vertical shafts. Numerous are the contrivances introduced for this purpose. At a few inclined shafts, sunk at high angle — above 65° — the trip-rope is used, but this necessitates the handling of the bucket, and also involves some danger, both to the top-man and to the miners below, in replacing the bucket on the skids. We have seen queer and ponderous arrangements hanging in the head-frame into which the bucket was pulled by the hoisting engineer, when a construction of steel bands and levers would automatically close upon the bucket and cause it to be overturned. At other places we have observed an ingenious contrivance which clutched the upper chime of the bucket and caused it to overturn when the bucket was pulled higher by the engine; but of all the numerous devices introduced for this purpose, we believe that here illustrated is the simplest and the best for all-round work. The design is furnished by Mr. Frank Robbins, E.M., of Los Angeles, California, who used it and found it to fulfil every requirement. Skips also are frequently employed in sinking shafts, particularly those of moderate dimensions, and it will be observed that there are points of material difference between the automatic-dumping bucket, and the automatic-dumping skip. In the former the bail is made fast to ears near the top of the bucket, while in the skip the side-bars extend entirely down the side to or near the lower end of the skip. The form of skip referred to will be more fully described and illustrated later.

The bucket here illustrated, Figs. 28 to 34, is intended to run on skids set at any angle which will permit the bucket to slide downward by gravity. If the inclination of the shaft is too flat to admit of this, due to friction of the bucket on the skids, use a skip with wheels. The several figures show the bucket in its various positions, both before and after dumping, and in the act of discharging its load into the chute *C*. The bucket is pro-

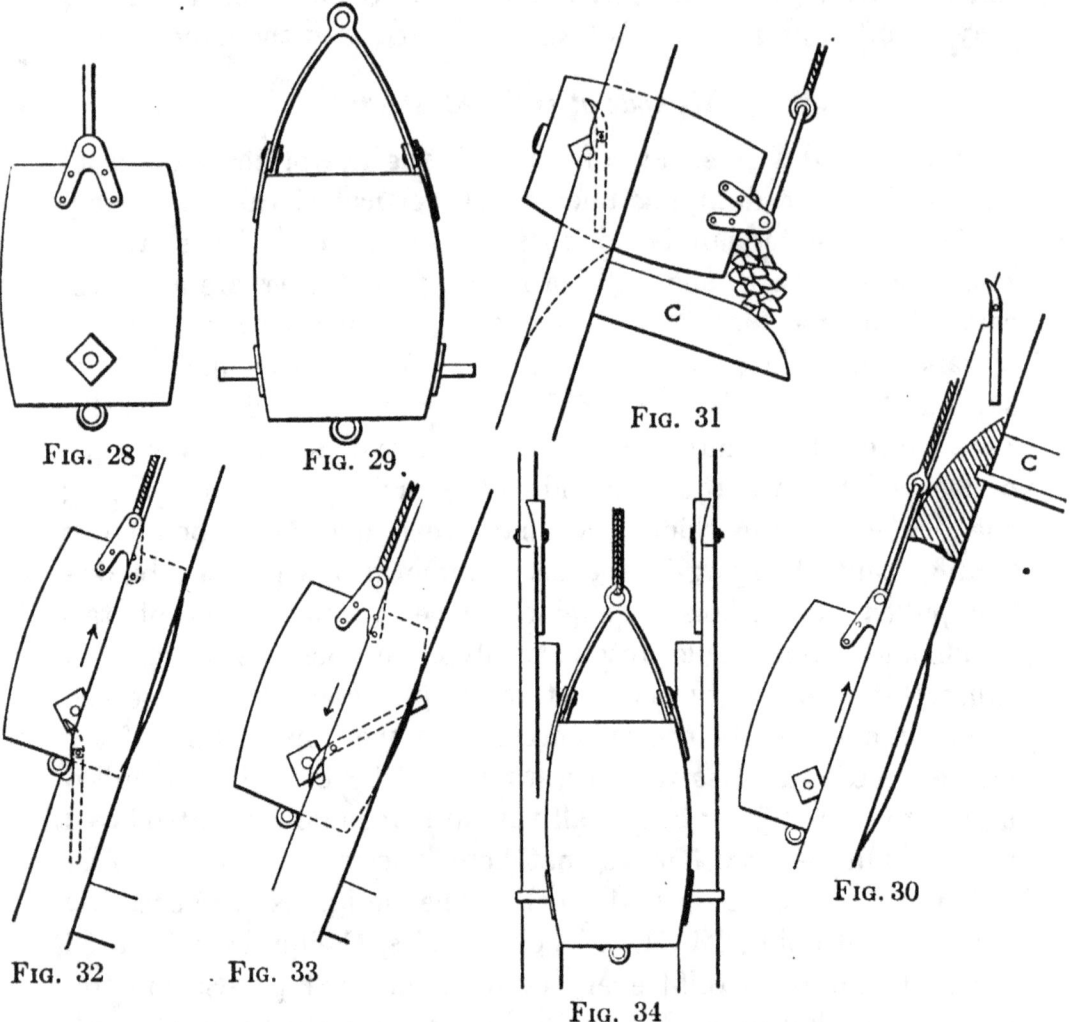

Fig. 28   Fig. 29   Fig. 31   Fig. 30
Fig. 32   Fig. 33   Fig. 34

vided with lugs, as shown in Figs. 28 and 29. The skids are placed at such a distance from each other that while ascending the shaft the lugs do not touch the skids. The lugs, as will be noticed, are considerably below the center of the bucket, and therefore below its center of gravity. Fig. 30 represents the bucket either in ascent or descent, sliding on the skids, the lugs being about 2 in. above the skids. At the dumping place the skids are channeled out on the inner edge, so that the bucket sinks low enough for the lugs to slide upon the skids. About 2 ft. higher a notch is cut in each of the skids, into which the lugs drop. The engineer then slowly lowers the bucket, which overturns as shown in Fig. 31. When empty, the engineer again hoists slowly, the bucket resumes an upright position, and is hoisted far enough for the lugs to be raised out of their notches and to pass above the latch attached to the inside of the skids, as shown in Figs. 30 to 34. On lowering again the lugs slip over the notches, as shown in Fig. 33, and the bucket again descends the shaft. In simplicity of construction and facility of operation this means of dumping buckets at inclined shafts certainly compares favorably with the methods previously described and illustrated for handling and dumping buckets at vertical shafts.

Where skips are used in place of buckets in sinking, in inclined shafts, the skips may be dumped automatically by turning the track rails from the angle of the shaft to a horizontal position. When the skip is hoisted from the depths of the mine, upon the forward wheels reaching the point where the track changes from an inclined to a horizontal direction, the forward wheels run out on the horizontal track, while the continued winding of the engine causes the lower wheels of the skip to be lifted from the track by the draw-bar, and the rock in the skip rushes out into the bin beneath. To accomplish this safely, a bumper must be placed on the horizontal track at the proper place (to be determined by experiment, its position varying with the length of the skip), against which the forward wheels will stop when hoisting. If desired, two or more dumping places may be provided, one above the other, the lower one being for a water dump, the upper one for ore and waste. When the skip is hoisting water a portion of the track is thrown back at the level of the water dump, and the skip will enter at that point and discharge the water. When ready to hoist rock, the sectional track, which is carried on hinges, is closed, and the skip, passing over these adjustable sections, dumps in the rock bin above.

At many inclined shafts the skips are provided with extension wheels on the rear axle, these wheels running on rails at the side of the main rails. This arrangement permits the forward wheels to follow over the curved portion of the track from the incline to the horizontal, the rear wheels taking the wider-gage extension track. The only advantage in this is that in lowering the skip after dumping there is no danger of the rear skip wheels striking the track a heavy blow. This a careful engineer can easily avoid where there are no extension rails, as we have seen more than a thousand times.

Where the angle of inclination of the shaft is very steep — approaching 90° — it is not uncommon to see the skip provided with a small wheel at one side, which engages a curved groove, the function of which is to guide the skip and to hold it steadily while dumping, and later in lowering again to position. Such shafts are usually provided with guides. (See Fig. 62.)

At vertical shafts dumping devices of the character last described are in common use and have been found to operate satisfactorily. At such places the groove should be used. If not, some precaution must be taken to prevent overwinding, otherwise a serious accident may result. When the horizontal track and bumper only are depended upon to dump a skip at a vertical shaft, the overwinding of a few inches will cause the forward wheels of the skip to run backward as the skip overturns, and a part of the load, at least, will fall into the shaft, which would be a serious matter to men below.

Chapter IX

FRAMING SHAFT TIMBERS

Before proceeding with the operation of shaft sinking, we will go somewhat into detail in the matter of framing shaft timbers. On this subject there is a most varied and widely scattered literature. It is not our purpose to give space to descriptions of, nor to illustrate, unusual methods, but to present what are generally recognized the world over as standard methods of framing shaft sets, for both vertical and inclined shafts. In the United States, in South Africa, and in Asutralia — wherever American engineers have been in charge — the rectangular shaft has been conceded to be the best adapted to all general purposes, though occasionally a square shaft is found. We have no particular interest in the great circular and octagonal shafts of Mexico and China, further than that they are rather remarkable examples of engineering of a somewhat primitive sort.

As to shafts sunk under peculiar and unusual conditions, such as are sunk under great physical disadvantages, special methods are employed — among them being sinking through quicksand by freezing the water-soaked ground; driving metal sheet-piling; sinking in caissons, and by the various other methods employed by engineers who make this class of work a specialty. Fig. 35 illustrates a type of shaft-timber frame at one time in much favor on the Comstock Lode at Virginia City, Nevada. It will be observed that the corners are framed so that the end plates do not rest upon the wall plates, but butt up against them. The idea of this design was to minimize the tendency of the timber to split under heavy pressure from the surrounding ground, and the attainment of this object was still further promoted by cutting a bevel at the inside corner of both end and wall plates, as clearly shown in the sketch, which is reproduced from Bulletin No. 2 of the State Mining Bureau of California, and which was drawn by the author in 1893.

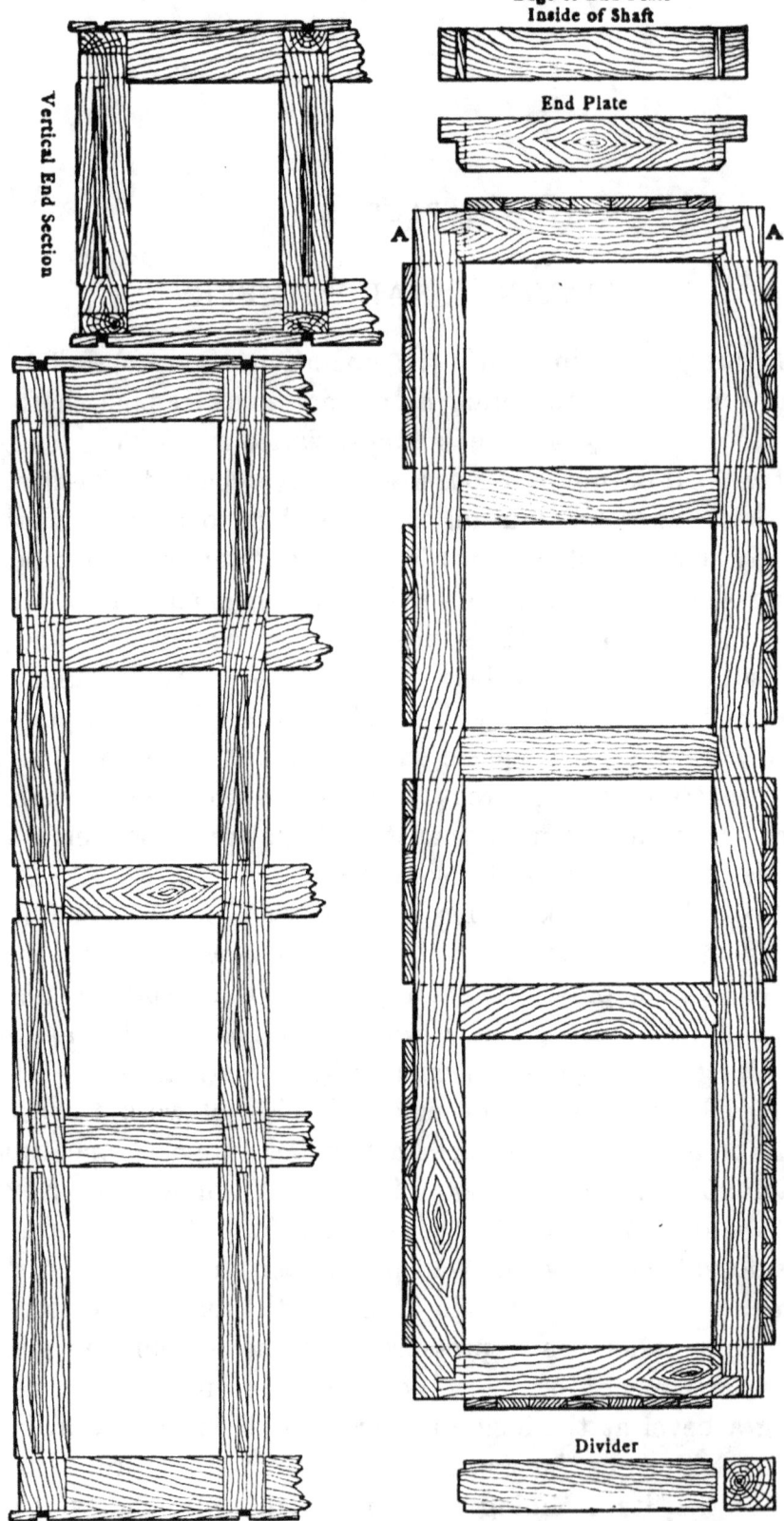

Fig. 35. — Method of Timbering the Chollar, Norcross and Savage Shaft. Comstock Lode, Virginia City, Nevada

Fig. 36 represents the shaft of the Argonaut mine, at Jackson, California. This is an inclined shaft, sunk at an angle of 63°, and is now nearly 3,000 ft. deep. It is a well-timbered shaft and an excellent example of the best type of inclined shaft in the West. The detail of framing is plainly shown. It has been suggested that the ends of the end plates be so framed as to dispense with the bevel at the outer edges, placing the end plates so that their outside corner comes flush with the extreme end of the wall plate. This would, of course, save a little time in framing, but the advantage gained, of holding the end plates in position until the set can be blocked and wedged, would be lost, and the present method is no small advantage in placing the timbers of the set in position; therefore we can see no advantage in changing the style of framing here presented for this class of shaft. Often in the smaller shafts, where comparatively small timbers are used — 8 × 8 or 8 × 10 in. — the outside bevel of the wall plate splits off when the end plates are put in. Two spikes driven into this part of the plate will prevent this wedge-shaped section from splitting off.

Fig. 37 illustrates a section of the Alma shaft at Jackson, California, and is an adaptation to vertical shafts of the "dovetail" style of framing employed in inclined shafts. It has been introduced extensively in the past ten years in vertical shafts throughout the West and appears to have fulfilled every requirement where the ground was not shifting by reason of unequal pressure or changing direction, and we are not at all sure that under such conditions the overlapping ends would afford any greater security, while we believe that timbers framed with the dovetail can be more readily repaired than those of the overlap style.

Framing for Vertical Shafts

Having described two methods of framing timbers for vertical shafts, a third method of framing is here illustrated. The sketch, Fig. 38, shows a shaft of two compartments designed for prospecting purposes; but as many compartments as are desired may be added, and the dimensions of timbers and size of compartments may be changed to suit the requirements. One or more compartments may be added by simply increasing the length of the wall plates to the necessary extent and inserting dividers at the proper places.

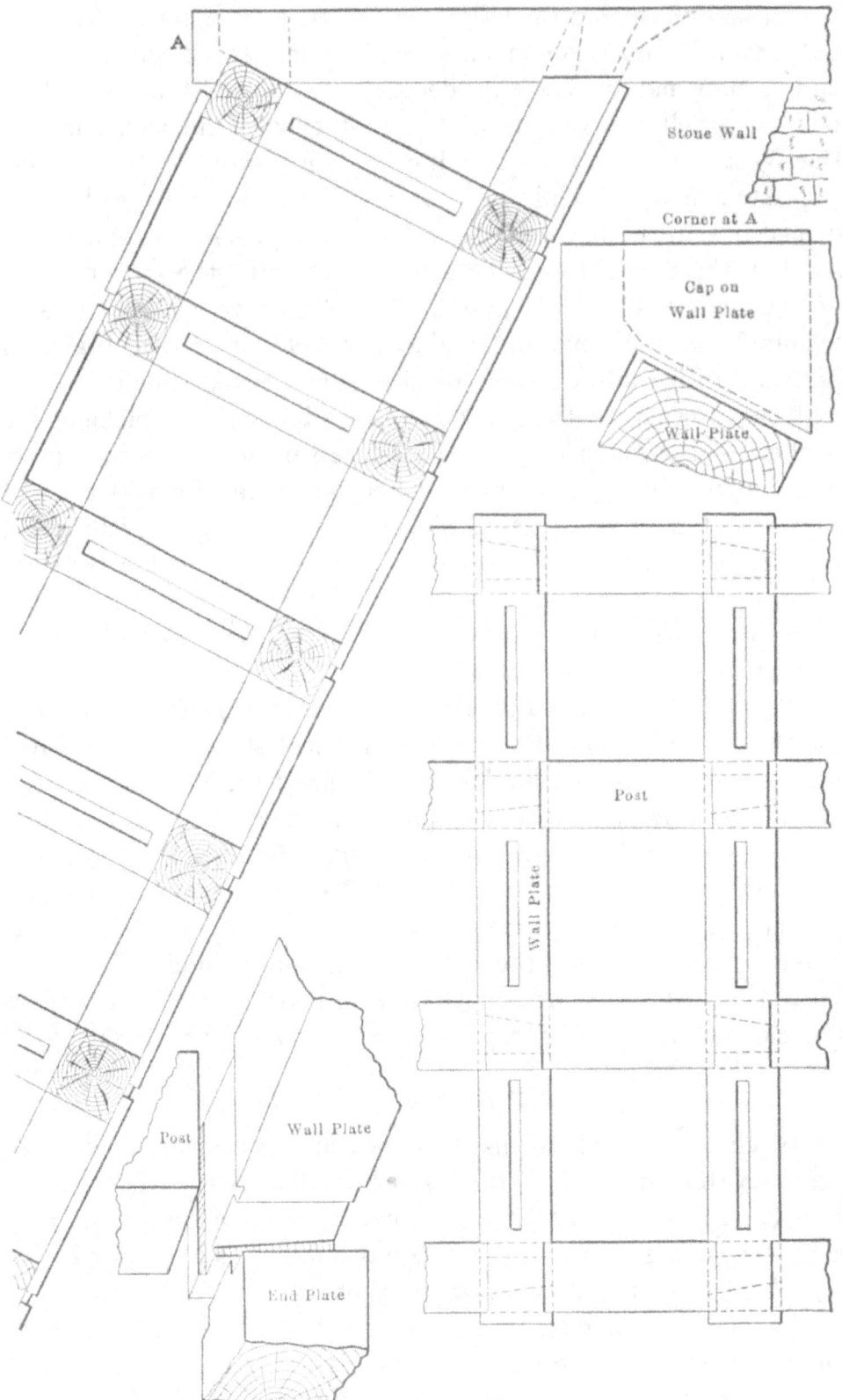

Fig. 36. — Argonaut Inclined Shaft, Jackson, California.

There is a detail of the framing of these timbers which we believe to be, in most instances if not in all, unnecessary, and

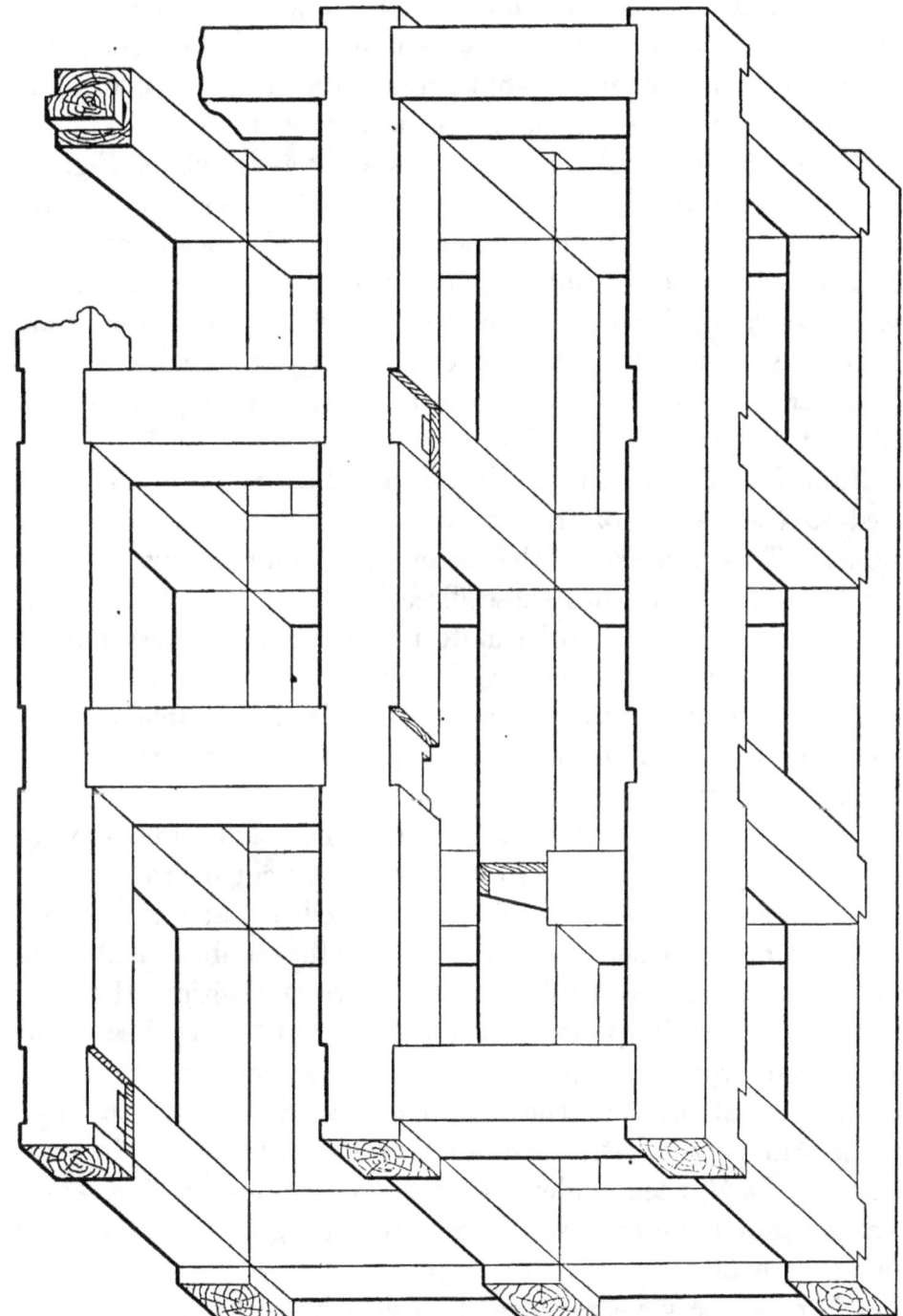

Fig. 37. — The Alma Vertical Shaft, Jackson, California

that is the bevel at the inside corner of wall and end plates. It is clearly shown in the drawing. The practice of framing timbers in this manner originated on the Comstock, the idea being that the bevel minimized the tendency of the timbers to split when exposed to heavy pressure; but it is our experience that when the rocks through which a shaft is sunk, swell, or from any other cause exert unusual pressure upon the timber sets, the tendency of the timbers to split and to crush is not materially decreased by the bevel at the inside corners of the set.

The sketch of the Chollar, Norcross and Savage shaft, Fig. 35, shows the same bevel at the corners as in Fig. 38. It requires time and care to make these bevels, and this means expense, which in our opinion is not justified by results. There is just as great a tendency for wall plates to split where the dividers are inserted, as at the corners, and even more, but no bevel joint has, so far as we are aware, been introduced at that point in any shaft, although there are a number of ways in which dividers are framed and inserted in shaft sets. Particular attention is called to the details of framing and instructions as to how to proceed. This appears in the lettering accompanying Fig. 38. It is very important that these directions be followed carefully, or the set will soon fail to be uniform, owing to the very noticeable discrepancy in dimensions of sticks of timber and an equal difference in the dimensions at opposite ends of the same timbers. This is due to the carelessness of sawyers in mills where the timbers are sawed from the logs.

In the early history of mining in the West, shaft timbers were held in position until wedged by the employment of iron dogs — bars of square iron having their ends turned out at right angles, and provided with sharp points. When the timbers had been adjusted as nearly as possible to their correct position, the dogs were driven into the plates and left there until several sets had been placed, when the upper set of dogs was removed to be re-used at the last set placed in the shaft. In some instances they were left in permanently. In some of the older shafts in California that had been under water for years, and had recently been unwatered, we have seen these interesting relics of the early mining methods.

It is now many years since the use of iron dogs was discontinued for that of bolts provided with a hook at one end and a thread at the other. At first a set consisted of one bolt having a hook and another an eye, but it was soon found that it was an advantage for each bolt to be made with a hook, and all of the

FRAMING SHAFT TIMBERS

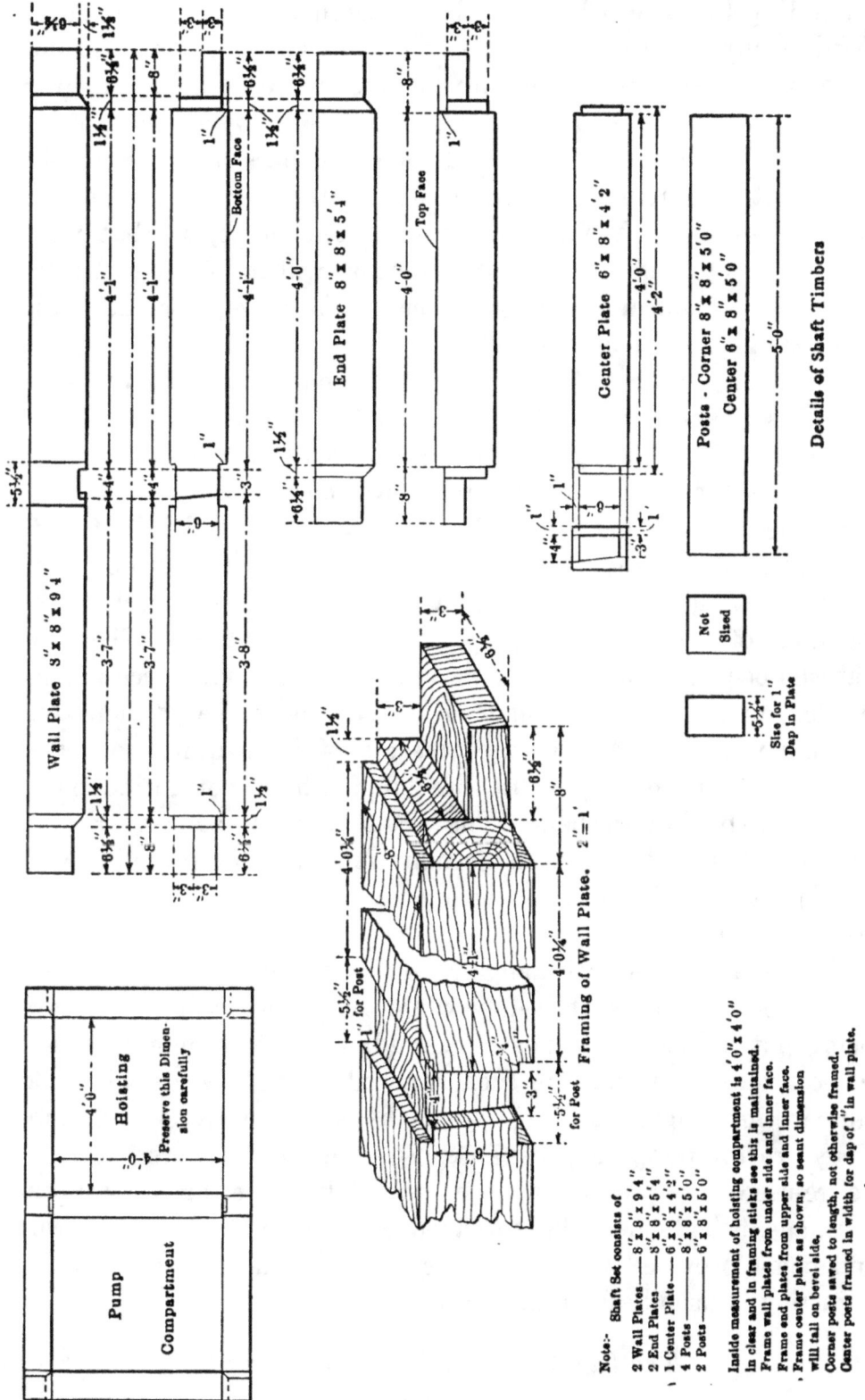

Fig. 38. — Details of Framing a Two-compartment Shaft Set

same length. The bolts were then interchangeable and could be used at any place in the shaft — either as an upper or lower bolt of a pair. We have heretofore made some remarks on hanging-bolts and wish here to repeat and emphasize some points of importance. Select good iron and have the bars of sufficient size to permit the cutting of a strong, deep thread, or the thread may strip. Forge the hooks into well-rounded shape, so that when two hooks are joined the point of tension will fall in line with the rods. Avoid burning the iron when in the forge, else the hooks will straighten out or break.

In the upper part of a shaft, or in any part of a shaft where the rock is soft and affords what is believed to be insufficient support to the timbers by wedging, it is a good idea to substitute straight iron bolts for the hooks and to allow them to remain in place permanently. In hard ground more than 5 or 6 full sets of bolts are not in use at one time, though it is wise to have two or more extra sets ready in the shop in case of an emergency, such as the breaking of a set, or the mislaying of a hook or two. When the threads become worn the end may be cut off, a new piece of iron welded on and a new thread cut. We repeat the caution concerning washers. Always use washers of large dimensions and sufficiently thick to prevent breaking, which not infrequently occurs with thin washers. The usual form of square or hexagonal nut is used, and at some mines the nuts are provided with handles to facilitate screwing up. By means of these much time may be saved.

The real function of hanging-bolts is to temporarily suspend the timbers in the shaft safely while all members of the set are being put in place, the posts inserted, and everything arranged in position. The nuts are then screwed up tightly, drawing the timbers into rigid position, when the set is wedged into line. They are for temporary use in most cases and are removed to be re-used lower down as the work of timbering progresses. Permanent bolts need be threaded at one end only, the lower end being provided with an "upset" head and a washer, as the adjusting is usually done at the upper end.

The Divider and Spliced Wall Plate

The dividers or centers placed in shaft sets for the purpose of dividing the shafts into two or more compartments are variously framed. Ordinarily the piece of timber is sawed to a length equaling the distance between wall plates, plus the length of the dovetailed or beveled tenon at each end. This latter may vary somewhat in different shafts, depending upon the size of the timbers of the set, a 12-in. or 14-in. plate usually being cut to a depth of 1¼ to 1½ in. to receive the tenon of the divider, while in an 8-in. or 10-in. timber the mortice and tenon are generally but 1 in. The tenons at the ends of the divider are always beveled, sometimes on both sides, as indicated by the dotted lines in Fig. 35, but more commonly one side only is beveled, the other being square, as shown in Fig. 38, in details of framing. In no case is the tenon flush with the side of the divider, the

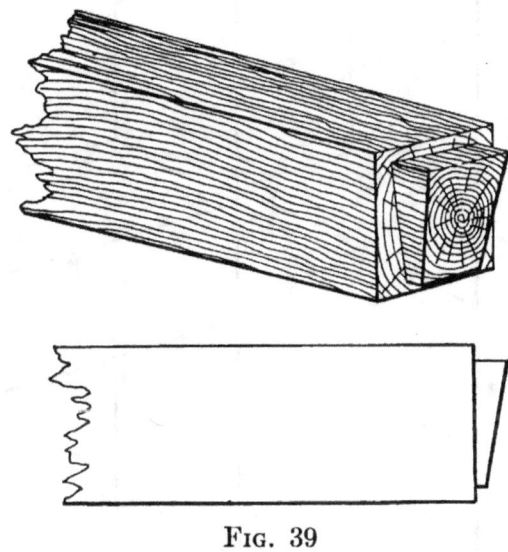

Fig. 39

object being to have the shoulders protect the wall plate as far as possible and to lessen the constant likelihood of its splitting under pressure.

Dividers should be heavy enough to perform the function for which they are intended — that of resisting side pressure upon the wall plates, and to carry guides and pipes in the shaft. It is customary, however, to provide for this purpose timbers which shall be equal to the inner face of the wall plate, but of less thickness. Thus a wall plate in a large shaft in heavy ground may be 12 × 14 in. The 14-in. face will be placed uppermost. The divider will then be 12 in. deep, but it may be only 10 or even 8 in. wide. In some shafts the dividers are square. As there is always a probability of dividers in the lower sets being knocked out of place by rocks flying from blasts, some miners cut the dividers with the tenon longer at the top than at the bottom, as shown in Fig. 39. As the "horn" projects further beneath the post, of course the danger of the divider being knocked out by blasting is lessened. Where there is heavy side pressure it is good practice to increase the height or depth of the divider, so that it projects an inch above and an inch below the surface of the wall plate, as shown in Fig 40, which prevents the center posts from being pushed out of place. This figure also shows a

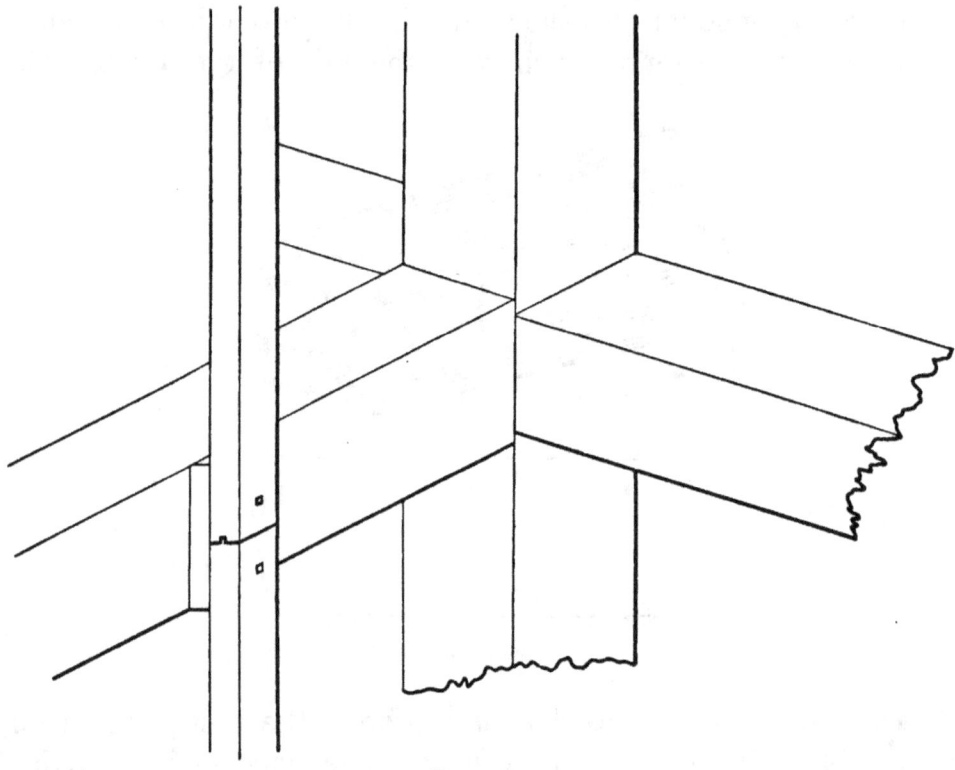

Fig. 40

guide in place, with what is known as a "distance piece" at the back of the guide, between it and the divider. The purpose of the distance piece is to enable the shaft men at all times to keep the guides in perfect alinement (within reasonable limitations) should the main members of the set be forced from their original position by unequal pressure. In heavy ground, where timbers frequently shift somewhat, the distance piece will be found particularly useful, as the guides can be kept in alinement without the necessity of shifting the main timbers until it becomes positively necessary.

Another idea in the matter of dividers is shown in Fig. 41. This scheme originated in Montana. The purpose is to make it possible to carry the main sets well toward the bottom of the shaft while leaving the dividers out of the last two or three sets so as to permit the swinging of long wall plates into position, and also to lessen the danger of having dividers knocked out by blasts. The square notch cut in the post at B permits the divider

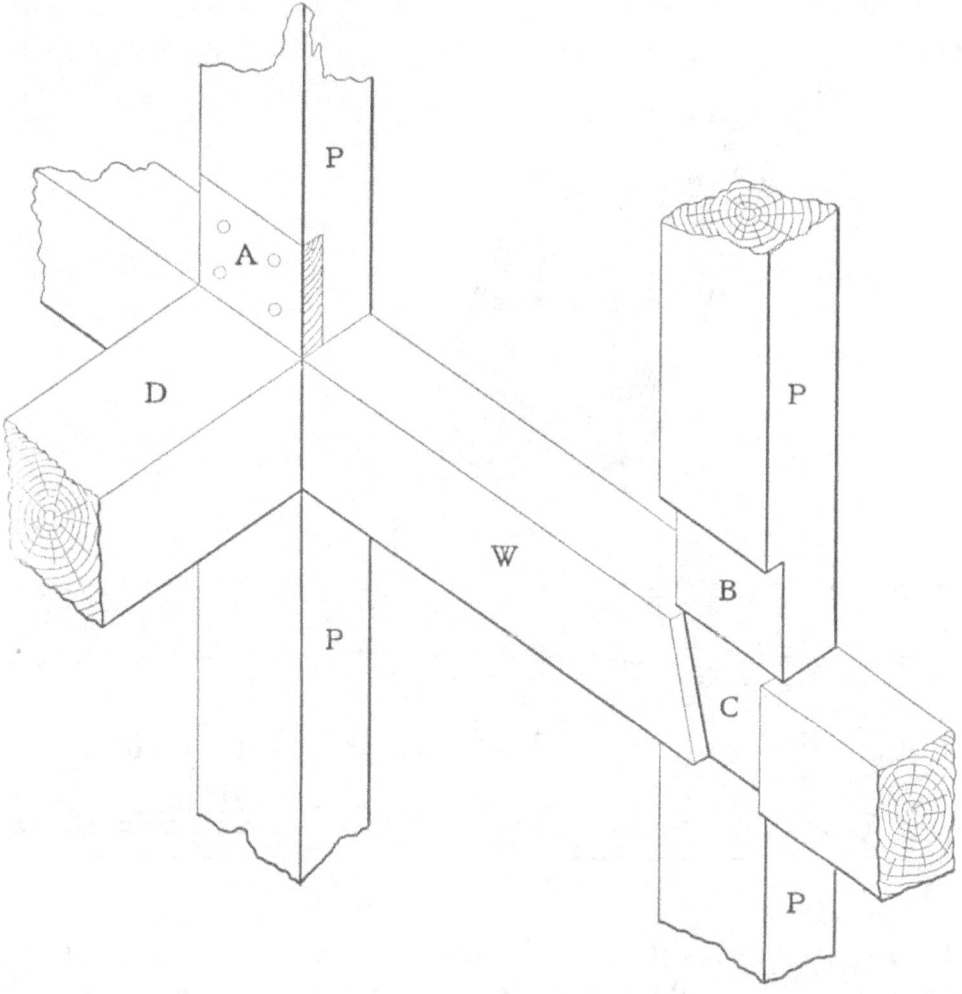

Fig. 41

to be slipped in sideways and then driven down to position, after which the notch at B is filled with a piece of plank which is secured by two or three small spikes, as at A. This scheme, although a convenient one, is not in general use, probably because miners believe the cut at A weakens the post; but as posts are only intended to afford stiffness to the set and to keep wall plates at the proper vertical distance from each other, it seems that the removal of a small section could work no great harm. If the piece of plank rots it can be removed and another substituted in its place.

Often it is difficult to obtain timbers of proper dimensions that are sufficiently long for wall plates in a large shaft, or it may be desired to enlarge a shaft by adding a compartment. In such cases the method of framing shown in Fig. 42 is a convenient and satisfactory way to meet this contingency. The ends of the wall plate butt against each other as shown, the

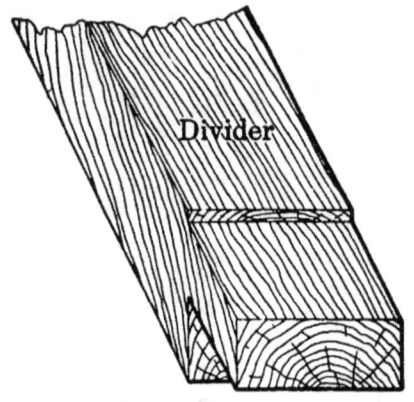

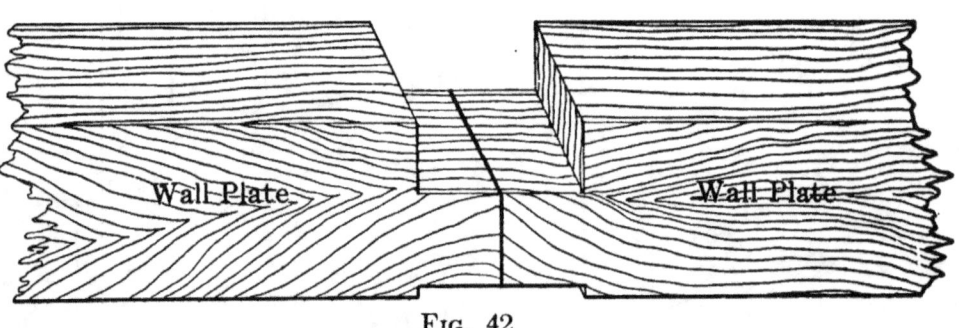

Fig. 42

divider, provided with a dap for the post, overlapping the spliced wall plate. This scheme can also be resorted to where the ground is so heavy that dividers cannot be omitted, but must be put in position at once, and all sets placed as soon as enough rock has been removed to make room for them. Manifestly it would be impossible to get a long wall plate into position under such circumstances, owing to lack of space within which to swing it into position. Then the spliced wall plate can be resorted to and the timbers carried as close to the bottom as desired. In such cases the two hoisting compartments are included by a single piece; the pumping and manway compartment being usually larger than either of the others, the short piece of the wall plate is placed at that end of the shaft. It suggests weakness, but when carefully framed, put in place and tightly wedged, a shaft set built on these lines is as good as any other. It is a mistake to think that shaft sets, or mine timbers of any kind, are improved by mortises and tenons such as carpenters commonly make in building houses, bridges, head-frames and other structures above ground. In mine timbers all depends on correct alinement of timbers and solid wedging. When these conditions are met they cannot be improved upon.

The Template

The framing of mine timbers, particularly those employed in shafts and in square-sets, must be done uniformly and with care. Timber-framing machines are now made that perform this work with expedition and accuracy. Of these we will have more to say later. Where shaft timbers are framed by hand, the work may be greatly expedited by employing a template, which may be laid on the timber and the marks scribed by the timber framer in charge of the work. The frame must be carefully made from a good piece of clear scantling, the templates being made of steel plates that are secured to the wooden strip by strong screws. The use of the template tends to insure uniformity in the shape and size of all cuts, and also keeps the distance from one to the other constant, whereas there is always possibility of an error through carelessness when each timber is laid out with scriber and rule.

Handling Timbers in Shafts

The handling of timbers at the collars of shafts, and the lowering of timbers into the shafts, requires great care on the part of those at the surface. The work must be done quickly, and absolute safety must be assured to the men below. At many inclined shafts a timber skip or car is provided upon which long pieces of timber, such as wall plates, long stulls, etc., may be sent down in safety. If there is sufficient room in the head-frame, the timber car may be attached beneath the ore skip, by rope or chains. It is good practice to have skids so arranged at the shaft that the timber truck may be run off onto the platform over the back of the shaft, where it may be pushed into the timber shop and loaded with timbers, which are then securely lashed on the truck and sent to the collar of the shaft; the rope or chains are then made fast in the ring in the bottom of the skip, and the timber truck with its load hauled across the shaft on the skids until the wheels rest upon the main track. The skids are then removed and the loaded timber truck and skip are lowered into the mine together.

Where there is insufficient height in the head-frame, the skip may be run onto the water-dump, where it is secured and, the hoisting rope being detached, made fast to the draw-bar or rings of the timber truck, and in that manner it may be lowered into the mine. If the timbers are for stopes, or for any other part of the mine than the shaft, a pair of skids should be provided at the stations similar to those on top, so that the timber truck may be run off at the level where the timbers are wanted. This will greatly lessen the danger to men who may be below, if the timbers are hoisted onto the level from the timber trucks by means of snatch-blocks suspended from the roof of the station. This danger is increased where the timbers are placed in skips or buckets, owing to the difficulty in removing them.

At vertical shafts the timber trolley cannot be used to advantage. Where cages are in use, however, timbers are generally stood on end and lashed to the frame of the cage. Long timbers, such as wall plates, project above the top of the cage, the steel hood being temporarily thrown back for this purpose. At most vertical shafts skips have replaced cages, though skips are not nearly as convenient as the latter for handling timbers; therefore with skips another means must be provided for the expeditious and safe handling of timbers. We have seen thousands of timbers 24 to 30 in. in diameter and 8 ft. in length, besides many of smaller dimensions, lowered beneath skips by means of chains and dogs. One of the mines where this method was followed was the Utica, at Angels, in Calaveras County, California. The stopes in that mine were large and the square-set system of timbering was employed almost exclusively; the timbers were mostly of the dimensions above given. These were delivered at the collar of the shaft ready framed from the machines. Only one timber at a time could be sent down beneath the skip. A dozen strong chains, each about 8 ft. in length, were provided. At the center of each chain was a ring 4 in. in diameter, and each end of the chain terminated in a strong, sharp dog, forged at an angle of about 70°. One of the dogs was driven into one side of a timber somewhat above its center, the chain being passed over the end, and the other dog driven into the opposite side. These dogs were driven well into the timber by means of sledges. They had to be absolutely secure, or a serious accident was likely to result.

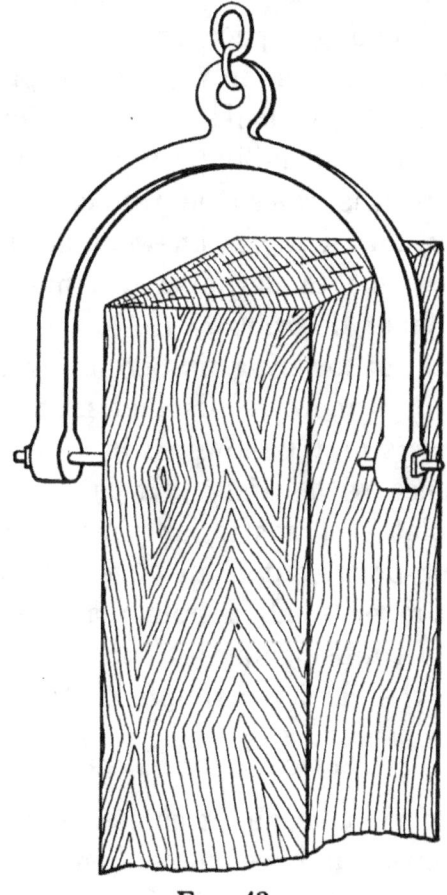

Fig. 43

In addition to the chains with dogs, a number of ropes 7 or 8 ft. in length, each provided with a sharp spike 8 in. long, were always on hand. The spike of one of these was driven into the lower end of the timber, and the stick was then ready to be lowered into the mine. When the skip came from below it discharged its load, and was then lowered to within about 5 ft. of the collar of the shaft. Beneath the skip was suspended two stout chains with rings. A short piece of chain having a strong hook at each end was at hand. One of the hooks was inserted into the rings at the ends of the chains hanging beneath the skip, the other end was caught into the ring at the center of the chain attached by the dogs to the timber. The signal being given, the engineer hoisted slowly, the timber being lifted and dragged toward the shaft, finally clearing the platform and swinging over the shaft. The top-man held the rope attached to the lower end of the timber by the spike, and by means of it brought the timber to a position of rest. The engineer then lowered the timber rapidly to the level at which it was to be removed. There the station tender pulled the hanging rope over to him, and with the help of several men landed the log on the station platform as the skip was lowered by signal. It looks like a tedious, expensive and cumbersome method of handling timbers, but with a crew of men accustomed to the work the timbers were handled with surprising rapidity, and I never knew of an accident occurring at the Utica mine due to a timber getting away while being lowered in the manner described.

A better method of lowering timbers is by means of a clevice — better because it is more secure and more expeditious. Fig. 43 shows how timbers may be lowered in a vertical shaft beneath skip, cage or bucket with safety. It is a method particularly adapted to wall plates of shafts, as these have holes bored to receive the hanging-bolts, but if an auger operated by power be provided in the timber shed, a hole may be quickly bored in every large piece of timber going into the mine, which will make the attachment of the clevice a simple matter.

Some Useful Knots

The handling of timbers about shafts necessitates the almost constant use of ropes, and we here illustrate the timber hitch (Fig. 44); the timber hitch and half-hitch combined (Fig. 45), and the bowline (Fig. 46). These three are the most commonly used by miners, and a knowledge of how to employ them will be found of great service. In making the timber hitch the rope is thrown around the timber and the loose end passed around the rope and over the loop, being merely wound two or three times around itself. The engraving shows the manner of making this hitch better than it can be described. Be sure to pass the end of the rope over and not under the loop. This is a very convenient hitch and suitable for lowering or hoisting timbers a short distance, but we would not advise its use in letting large timbers down a shaft, as sometimes timbers slip in the timber hitch, particularly round sticks, and this tendency is greatly increased if the timbers are wet. For greater security a half-hitch may be added as shown in Fig. 45. This takes a double hold upon the timber and rarely slips. The half-hitch should be at least three times the diameter of the timber from the timber hitch, as this further decreases the likelihood of the rope's slipping.

The bowline is probably the most useful knot known to miners, and is borrowed from the sailors. It was doubtless introduced

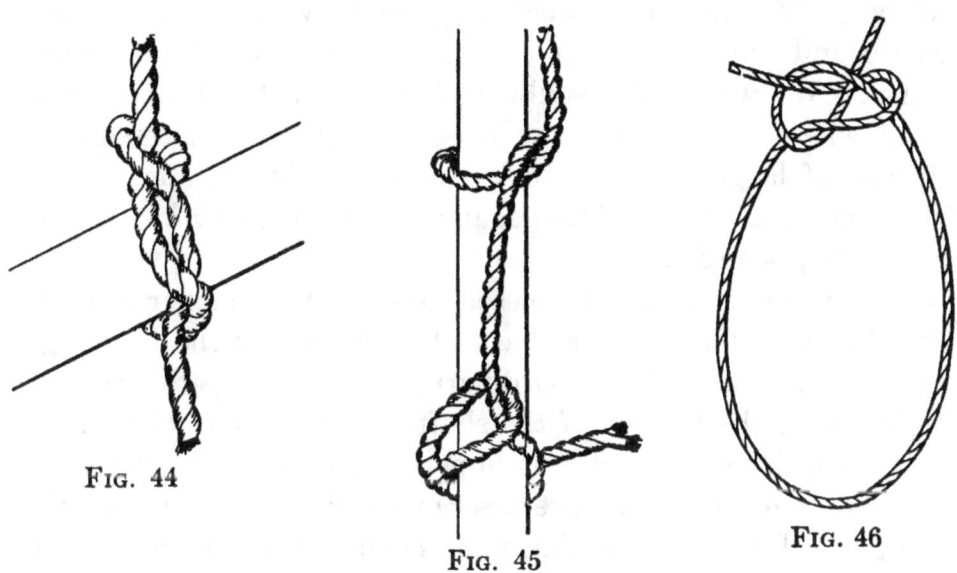

Fig. 44   Fig. 45   Fig. 46

into mining practice by Cornish miners many years since, as a great many of them are as able seamen as they are miners. It is difficult to describe the method of making the bowline, but the engraving shows it so plainly that no further instruction should be required. Take a piece of rope and practice throwing the bowline. It is a trick that can be quickly acquired and will be found useful a thousand times. It is a knot, quickly made, never slips when properly made, and is easily untied. These three knots will be found most useful in handling timbers, and are more employed about vertical than about inclined shafts.

Placing Timbers of Shaft Sets in Position

Considerable space has been devoted to the framing of shaft timbers, for both inclined and vertical shafts, and something has been said about the means employed in lowering heavy timbers in shafts, whether these are to be used at the bottom of a shaft in process of sinking, in repairing the shaft, or to be taken off at one of the levels for use in a stope or elsewhere. A great deal more could be said, perhaps, concerning the framing of shaft timbers, but such remarks would necessarily be confined to shafts of special size and appointments. Such shafts are usually sunk under the direction of an engineer who has planned the shaft and provided all of the special arrangements for that particular equipment. However, it is our purpose to present one or two more designs in vertical shafts showing particular appointments that have been suggested by necessity and experience and have been found of great value in the operation of mines. We will also describe and illustrate the timbering of a shaft where the ground is so loose or soft as to necessitate the driving of lagging as the work proceeds. This is done in much the same manner as where lagging is driven in drifts, the principle being identical.

Before proceeding with these, however, the manner of putting the timber in place in the shaft when it reaches the timber gang below will be described. Ordinarily the timber sets are carried well toward the bottom of the shaft as the work of blasting proceeds, and in some cases it is imperative to keep the timber well down to the bottom to avert serious caves. This is sometimes the case where no forepoling is necessary, but wherever possible it is better to carry the work of blasting well in advance of timbering, for several reasons, among them being the freedom from damage to timbers from flying rocks, and the fact that drilling may proceed below while the timber gang is at work above, thus saving much time, which is often an important matter. In the shafts of the Rand, South Africa, it it not at all uncommon for the bottom of the shaft to be from 100 to 150 ft. in advance of the timbering, the skip or bucket being held in position below the timbers by a cross-head running in wire-rope guides. When the bottom of the shaft is so far below the last set of timbers that men cannot place the timbers while standing on the bottom of the shaft, the first thing to be done is to place two or more stulls (depending on the size of the shaft) across the opening. These must be placed with great care, and should preferably be set in hitches cut in the rock and then tightly wedged. Much depends on the absolute security of these stulls, upon which a platform of planks is to be laid, for not only must it sustain the weight of planks and several men, but also, possibly, a ton or more of timbers, tools, etc., before the set is in place. On the security of this platform depend the lives of the timbermen, and of the men at work below, should there be any. The platform must be made tight in case men are below, so that no tools, blocks, or even the wedges can fall through. When the platform is in place — usually about 7 ft. below the last set, the first wall plate is lowered, as previously explained, the last two or three dividers having been omitted from the sets above in order to permit space to handle the long wall plates when they arrive from the surface. As the heavy timber comes within reach of the men the lower end is carried toward one end of the shaft, and as it descends it finally is laid flat on the platform. The engineer stops the skip or bucket upon signal. The rope, or clevice, is detached, and a rope passed around the wall plate and made fast by means of a securely arranged timber hitch. The upper hanging-bolts have already been put in place in the lowest plates of the set above. The hooks are now placed in the wall plate about to be hung up. A glance at the framing will show to which side of the shaft the plate is to go. Generally it is understood with the top-man which plate is to be sent down first. This prearrangement may obviate the necessity of turning a plate end for end and dispense with much profanity. The rope is now made fast to the ring in the bottom of the bucket by means of the bowline. All being in readiness, the signal is given the engineer to hoist slowly — generally 3 — 1. He knows what is going on below from experience and slowly starts the engine. In a moment the rope is taut and the wall plate is lifted and is nearly balanced. The men guide it and push it gently to the side until the hooks in the new plate can be engaged in those hanging from that above. When this is accomplished, the signal is given to stop hoisting — 1 bell. The engineer then lowers on two bells; the rope is detached from the timber and the bucket rung up. The second wall plate is sent down and placed on the opposite side of the shaft in exactly the same manner. The ends may then be sent down, or may already have been sent down, together with the posts. These are placed on the platform.

If the shaft set be framed with the dovetail, it will be found convenient to have a gage to place between the wall plates to adjust them exactly to the necessary distance apart. This distance is determined at each end and the plates held in position temporarily by spiking a piece of lagging to the top, using small spikes. This is done at each end. The end plates are then dropped into position, the bevel keeping them from falling through, and the spiked lagging preventing the wall plates from spreading.

The posts are now placed on the corners and in the proper daps along the sides. In some shafts posts (studdles) are used only at the corners, but if great rigidity is desired, posts must also be placed opposite each divider. When the posts are in position the hanging-bolts are screwed up tightly, and if the daps are cut to the proper depth and the posts are of uniform length (as they should be for every set) the timbers will come into exact position and the plates will be level. This must be constantly watched in all shafts, whether inclined or vertical. Should the plates get out of level from pressure or other causes, they must be brought back to level within a few sets, without in the least disturbing the alinement of the shaft. The writer knows of an inclined shaft which, through careless framing and wedging, got out of level. The foreman undertook to remedy the difficulty by cutting the posts short on one side for three sets. He neglected the alinement and in the next 200 ft. the shaft had taken a "twist," so that in that distance it traveled north from the correct alinement nearly 13 ft., and was twisted around so that had it continued 1000 ft. further in the same way it would have resembled a gigantic corkscrew. This was about as bad a job as we have ever seen, but even an Italian foreman should have sense enough to keep the hoisting rope in the middle of the track, though this particular one seemingly did not.

If the overlap framing is in use the end plates may be laid upon the ends of the wall plates and the temporary binders may be dispensed with by immediately placing blocks at the corners and wedging the set up firmly but not too tightly, for the shaft has yet to be lined up accurately. When the posts have all been placed, whether in vertical or inclined shaft, the set should be blocked and wedged to the rocks surrounding the shaft.

FRAMING SHAFT TIMBERS

Placing Timbers in Inclined Shafts and Lining the Sets

The placing of timbers in inclined shafts is attended with more difficulty than in performing the same work in vertical shafts. If the angle of inclination is less than 70°, and the timbers large in size, much lifting is necessary. In general practice the timbering is kept well down toward the bottom as work of sinking advances. This is in part due to the increased difficulty in placing the timbers in position, and in part to the greater danger of the caving of the hanging-wall of the shaft.

It is at once apparent that in an inclined shaft, unless it be at an angle approaching 90°, the heavy wall plates cannot be put in position in the same manner as the plates are placed in vertical shafts. There is always more or less lifting, and the men of the timber gang must be strong and experienced in the work. When a new set is to be placed at the bottom of an inclined shaft, the first thing to be done is to ascertain that the ground is clear of the outside alinement of the set on all sides — that there are no projecting points or shoulders which may later prove to be in the way and have to be beaten or moiled off. This kind of work can be done much more quickly when no timbers interfere with the work. It is particularly necessary to see that the foot side is clear. The sill plate is the first sent down from the surface. This is placed in position with great care. It must be in alinement with the sets above and must be perfectly level. This latter point is of greatest importance, for if the foot-wall plates get out of level the shaft is soon in bad condition, and it may be necessary to do much expensive work to bring the timbers into exact line. A mistake in side alinement would, in any event, probably be so small as not to be quickly noticeable, and may usually be corrected by wedging without much trouble. Even a mistake in angle of inclination is not so serious, for if too high the plate can be cut down with chisels to lower the rails to the proper height; or if too low the plate may be wedged or built up; but if not level, great difficulty is likely to be experienced in remedying this defect.

The three things, then, to be most carefully observed are that the foot-wall plate is absolutely level — not higher at one end than at the other; that it is exactly in line with the angle of inclination of the shaft, unless raised or lowered intentionally, where the angle of inclination of the shaft is to be changed; and to further see that the direct alinement of the shaft is maintained. All of these things can best be done by means of a transit, but as mine foremen are not all familiar with the use of the transit, and as every mine does not employ an engineer, other less refined methods are often employed that give very satisfactory results.

Among the tools generally employed for the purpose are the spirit-level, steel-square, and a 20-ft. straight-edge. The level is indispensable in placing the foot-wall plate in absolute horizontal position, and the square is useful in placing the end-plates vertical in reference to the side. The straight-edge is looked upon as an instrument of precision by the shaft men, but it is often something else. It is subjected to more or less abuse, receiving many a blow, is exposed to heat and cold, wet and dry, and if it does not warp a little it is a wonder. The edges in time become worn unequally, and though as perfect as the ingenuity and skill of the carpenter can make it when new, it in time, through these many abuses, becomes unreliable, and a slight error due to a defective straight-edge is repeated and perpetuated until the shaft, in a distance of 200 ft. or more, will exhibit a beautifully true transition curve. An attempt, too late, is made to correct this, and an abrupt bend is made, not noticeable until a dozen sets are in place, when it may be seen plainly enough on looking along the timbers, if there be light enough to see it.

One of the best ways to secure alinement is to make a mark or, better, to drive a tack about 6 in. from the inside corner at each end of the plate. Some make a light cut with a saw, but this is bad practice as it weakens the timber, and under pressure this weakness is noticeable, for the timber frequently begins to split at the "saw-craft." To drive a bright tack — galvanized tacks are good — is a better scheme. These tacks, in the center of the timber and 6 in. from the end plate, are always easy to find. Then, when four or five sets have been put in place under the direction of an engineer or some one familiar with the use of the transit, the work may proceed by employing so simple a thing as a tightly drawn chalk-line: one man five or six sets above; one at the new set (or the line may be made fast at the proper places by means of two nails) and the foreman, or timber boss, along the line, particularly at the last completed set. If the new plate on the bottom is not in alinement, the chalk line, held or nailed exactly over the tack in the new plate, and over that on the sixth plate above, will not cover the tack on the last completed set. The new plate must then be driven by means of wedges or otherwise to right or left until the lines do coincide. For the sake of greater exactness, and proof, the line should be tried on the tacks at the opposite end. The spirit-level will, if employed with care, be sufficient in securing the plate in horizontal position. When the foot-wall plate has been placed and wedged at the ends so that it cannot shift, the hanging-bolts should be put in, and the studdles, or posts, set in their respective daps and the bolts screwed up.

The men are now ready for the hanging-wall plate. This comes down with the galvanized tacks driven on the inner side at exactly the same distance from the corners as in the foot-wall plate. When ends and all the posts are at the bottom of the shaft, the timbers are distributed so as to have them handy. If not too heavy the men lift the hanging-wall plate and the ends are slipped in from above; the upper studdles are placed in position and held there, while men adjust the hanging-bolts of the upper plate and screw them up as quickly as possible. The set is then wedged temporarily, and preparations made to proceed with the alinement.

If the wall plates are long and very heavy, the labor may be lightened somewhat by first placing two skids reaching from the foot wall of the shaft upward to the rock face and below the position to be occupied by the plate. The plate is then sent down and is laid across the skids, when it may be rolled or slid upward on the skids to a position approximating the one it is to occupy permanently. The ends may then be slipped in, the hanging-bolts inserted, and the posts placed, when the bolts may be tightened. When the set is in place the skids may be removed, they having served their purpose. Blocking is then put back of the timbers at the corners, or wherever necessary, and wedged, but not too tightly.

The foot-wall plate being in absolutely exact position the hanging-wall plate must now be brought into line. Chalk lines are made fast on the new hanging-wall plate, and at the seventh hanging-wall plate above, and the lines are also replaced on the foot-wall plates. Plumb-bobs with short lines are attached to the upper lines at a point vertically above the set number 2 (calling the new set number 1). The plummets should be of the long kind terminating in a sharp point. This point should hang directly over the lower line, for if it does not the ends are not in exact position and the hanging-wall plate must be shifted by easing and wedging until correct adjustment is secured. This may also be done by means of the spirit-level held upright, but there is more likelihood of error resulting from inequalities of the timber and short distance than by means of the lines and plumb-bobs.

Difficulties in Sinking Through Running Ground

Sometimes shafts in sinking must pass through soft, water-soaked rock. Whenever this occurs, no matter what the kind of rock, whether the muck of a slaty gouge, the soft unctious talcose material of talc-schist or crushed serpentine, or whether it be "rotten" granite or crumbling sandstone, there is trouble in store for the miner, and often danger as well. Decomposed sandstone and limestone will sometimes run like sand when dry, and this tendency is increased to a dangerous degree when the country is wet. The experience at the Alpha shaft of the Giroux Consolidated mine, at Ely, Nevada, in 1908, was an example of the dangers of sinking through loose material. A run started in that shaft at some unprotected place, and this loosened some of the blocking, which permitted the lagging of a set to drop out, and the shaft simply collapsed for hundreds of feet, as the soft disintegrated material shifted, loosening the blocking. It cost the lives of two men and many thousands of dollars, and it is a wonder that the cave was ever caught up at all. Great praise is due the men who heroically worked in imminent danger for weeks to rescue their companions imprisoned below.

Leadville, Colorado, has numerous instances of caving shafts, caused by stoping ground too close to the shaft. Of these the Bon Air shaft is an example. Hundreds of bales of hay were dumped into that hole and the run was finally checked and the shaft recovered. But these are instances of caving and recovery after the shafts had been put down.

There are many places where shafts may be sunk in safety through soft ground, so long as it is dry, but let water come in contact with it and nothing short of several feet of concrete could hold it. There are numerous examples of this sort of ground in the mines of the Mother Lode of California, of which the Amador Queen, No. 1, a mile and a half south of Jackson, in Amador County, is a type. The shaft is over 1200 feet deep. Heavy gouges are an important geological characteristic of this mine, but it is dry and no great difficulty was experienced in sinking a good-sized two-compartment shaft at a cost of only $30 per foot, which on the Mother Lode may be considered as indicating "easy ground."

Where a shaft must pass through water-soaked running ground extraordinary precautions must be taken. Some mining engineers have made a specialty of sinking shafts through quicksand and wet, running ground, and in some instances elaborate and expensive preparations had to be made and special methods employed. Among these are the freezing of the ground by means of pipes driven into the ground to be excavated, and circulating ammonia through the pipes, operating in essentially the same manner as ice-making machines. This causes the ground to congeal for a time, when it may be excavated, and the excavation made secure by means of steel sections or massive timbers surrounded by tight lagging. This process is exactly the reverse of shaft sinking in artic regions where the ground is solidly frozen and must be thawed by means of pipes filled with steam under pressure from a boiler and driven into it as the ground softens.

Sinking Through Running or Loose Ground

As previously stated, the sinking of shafts through running ground, or that which is loose and caves readily, is attended with both difficulty and danger. To overcome these difficulties various special methods have been introduced with greater or less success, some of them expensive and elaborate, but the method most commonly employed is that of driving lagging (forepoling) into the loose ground, thus forming a tight box-like dam around the outside of the timber frame, inside of which the work may progress with safety and with greater or less expedition, dependent largely upon the conditions to be met, chiefly the character of material to be removed and the amount of water present. Where water is abundant in loose material, particularly that which runs readily, it is a good maxim to "make haste slowly," for an attempt to rush the work may meet with a back-set. It is much better to go about it deliberately, giving the ground time to drain, by cutting a sump hole at some convenient place in the bottom of the shaft and removing the water as fast as it comes in, either by means of a sinking-pump or by bailing, the means to be determined by the amount of water coming in.

It is very necessary to have first-class timbers and lagging for sinking through ground of this sort, and while split spruce lagging may be preferred in passing through dry ground, or where the rock stands fairly well and will not run, smooth sawed lagging is absolutely necessary in this kind of ground, because split lagging has many inequalities of surface, and there is often considerable space between adjacent pieces, due to these inequalities. The sawed lagging must be placed close together and the swelling of the wood due to the water will cause all inequalities to close, and if properly placed and sufficiently strong they will permit sinking through very bad loose and running ground.

When sufficient progress has been made in the shaft to admit of placing an additional set of timbers, the main members of the set — wall and end plates, posts, dividers, etc. — are placed in position in the manner already explained in former chapters. There is this important difference, however. In the sets now under contemplation a "bridge" is used. This is a piece of timber of a size to be determined by the size of the shaft, dimensions of main timbers, and condition of the ground. Ordinarily timber 4×10 in. is heavy enough. This bridge is in a single section at the ends of the shaft, but may be made in two or more sections, at the sides. These are placed outside the wall and end plates parallel with them and separated from them by wedge-shaped blocks, as shown in Fig. 47, where B is the bridge and W the wedge blocks. The end of the lagging is inserted between the bridge and the wall plate and driven downward as the work of excavation progresses, and ahead of excavation, the ends being driven into the soft ground as far as possible. The lagging is kept pointed outward by means of a block at C, which prevents the upper end from working outward as the lower end meets the resistance and pressure of the ground below. In some cases it may even be necessary to employ the false-set, though this is not of frequent occurrence.

In some large shafts a small shaft is sunk within the main shaft, something after the manner of digging sewer trenches in city streets. The chief object of this is to carry a sump in advance which will relieve the ground of much water, rendering it less troublesome to cut out the ground for the main shaft, and to place the timbers with as little delay as possible. Lagging is driven on all four sides of the shaft, always as close as possible, and the work is advanced on each side evenly — that is, one side or end must not be much in advance of any other. The process of sinking shafts under the conditions here contemplated is very much the same in principle as driving lagging in drifting through similar ground, and the methods of drifting are applied to shaft sinking in essentially the same manner, even to the carrying of breast-boards, which has previously been described and illustrated.

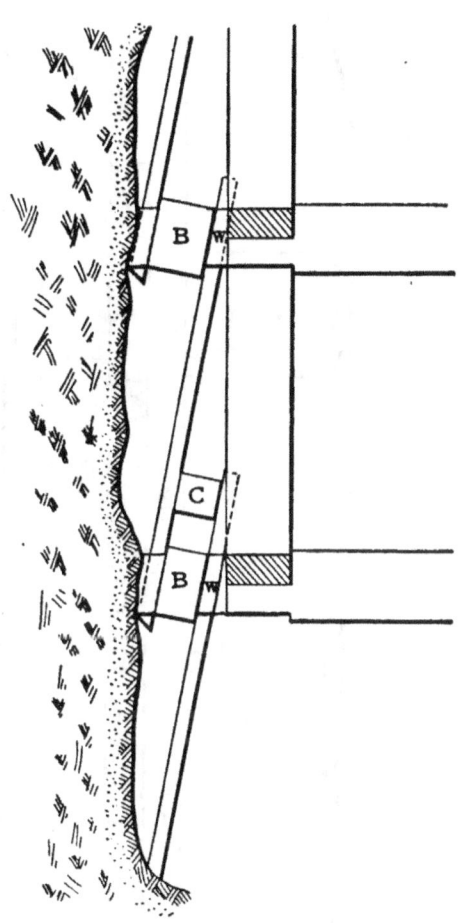

Fig. 47

No hard and fast rules can be laid down for work of this character. The workmen must have experience, or gain experience in trying to do the work themselves, and in timbering shafts experience is a most excellent teacher. The fundamental principles have been clearly stated, and the work or method has been illustrated. The timberman must necessarily have some initiative himself, and act promptly should anything unlooked for occur. The first thing to be considered is safety to the men employed; the next, expedition in the work, and this will be determined by conditions, and as has already been said, the greatest speed is not always attained by rushing a shaft through soft, water-soaked ground.

The east shaft of the Kennedy mine at Jackson, California, is to-day the deepest vertical shaft in California, and one of the deep shafts of the world, being now more than 3100 ft. deep. This

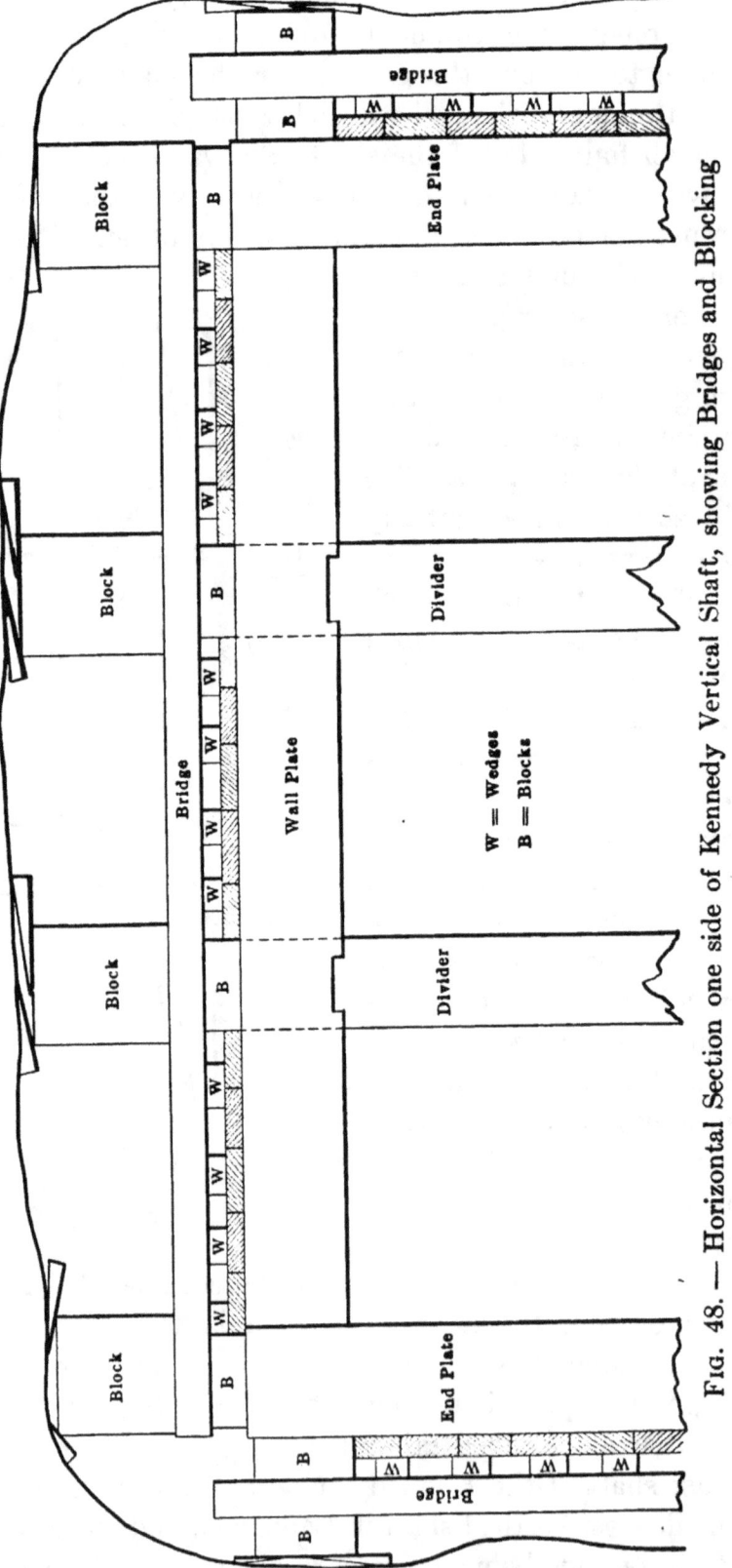

Fig. 48. — Horizontal Section one side of Kennedy Vertical Shaft, showing Bridges and Blocking

shaft is sunk in hard greenstone, and no pressure developed nor need have been anticipated in such ground, but when the upper part of the shaft was sunk in 1900, it was deemed advisable to employ the bridge. Just what object was in view we do not know, unless it was thought that by the employment of this method it might be easier to keep the shaft in line, in case of subsequent movement of the rock walls — of which there was not the slightest likelihood. Possibly the idea originated from the experience of the next mine on the north — the Oneida,

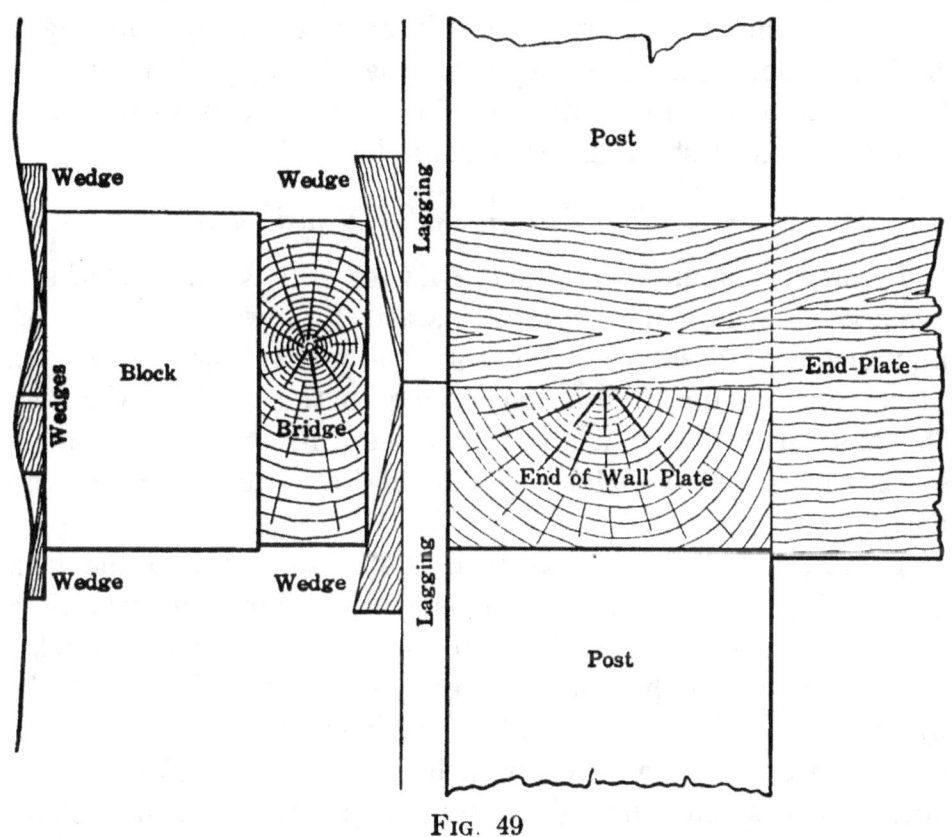

Fig. 49

where a vertical shaft was sunk through hard greenstone, and which encountered the vein fissure at 1900 ft., where the rock was black slate on both walls with a heavy gouge. The movement of the rocks in the vicinity of the vein afterward caused the Oneida management much trouble, and possibly it was to avoid similar difficulty that the Kennedy management adopted the bridge early in the work. This was wholly unnecessary, for the bridge could have been introduced or discontinued at any part of the shaft, and moreover, when a shaft passes out of hard ground into soft, swelling ground a different and more substantial method of proceeding than the introduction of the bridge should be resorted to. The bridge is useful in passing through running or very loose ground, but is not calculated to resist the pressure due to swelling ground.

The accompanying illustrations, Figs. 48 and 49, show the plan and end at one corner of the Kennedy shaft. It will be observed that it differs materially from the arrangement shown in Fig. 47. The bridge in the Kennedy shaft is without usefulness, and therefore an unnecessary expense, which might have been omitted until it became a necessity. It will be noticed by referring to the illustrations that the lagging in the Kennedy shaft was not driven, but placed in position in the same manner as lagging is usually placed in shafts in hard ground.

Combination Vertical and Inclined Shafts

In many instances it is necessary to sink vertical shafts to reach veins which lie at a comparatively low angle, and not infrequently it is advisable to turn the shaft from the vertical direction to an inclination conforming to the pitch of the vein. This has been done at a number of deep mines on the Rand in South Africa, on the Mother Lode of California, and in other places.

In numerous instances vertical shafts are sunk to the vein, and, passing through it, the shaft is continued in a vertical direction, the vein being reached by cross-cuts, both above and below the level at which the vein is intersected by the shaft; but this practice is, or should be, confined to veins having a relatively high angle of dip, those of low inclination being worked, as previously mentioned, through a combined vertical and inclined shaft, where, as on the Rand, it is necessary to reach the vein through a vertical shaft, due to the fact that the outcrop is owned by others.

There are important considerations that cannot be overlooked when contemplating this change from vertical to inclined, particularly where rapid running is necessary to hoist large tonnage within limited time. In such cases it is inexpedient, or at least undesirable, to slow up the skip when running past the curve, and for this reason the work must be directed with care by an engineer, who will lay out a transition curve in such lines as to avoid as

FRAMING SHAFT TIMBERS

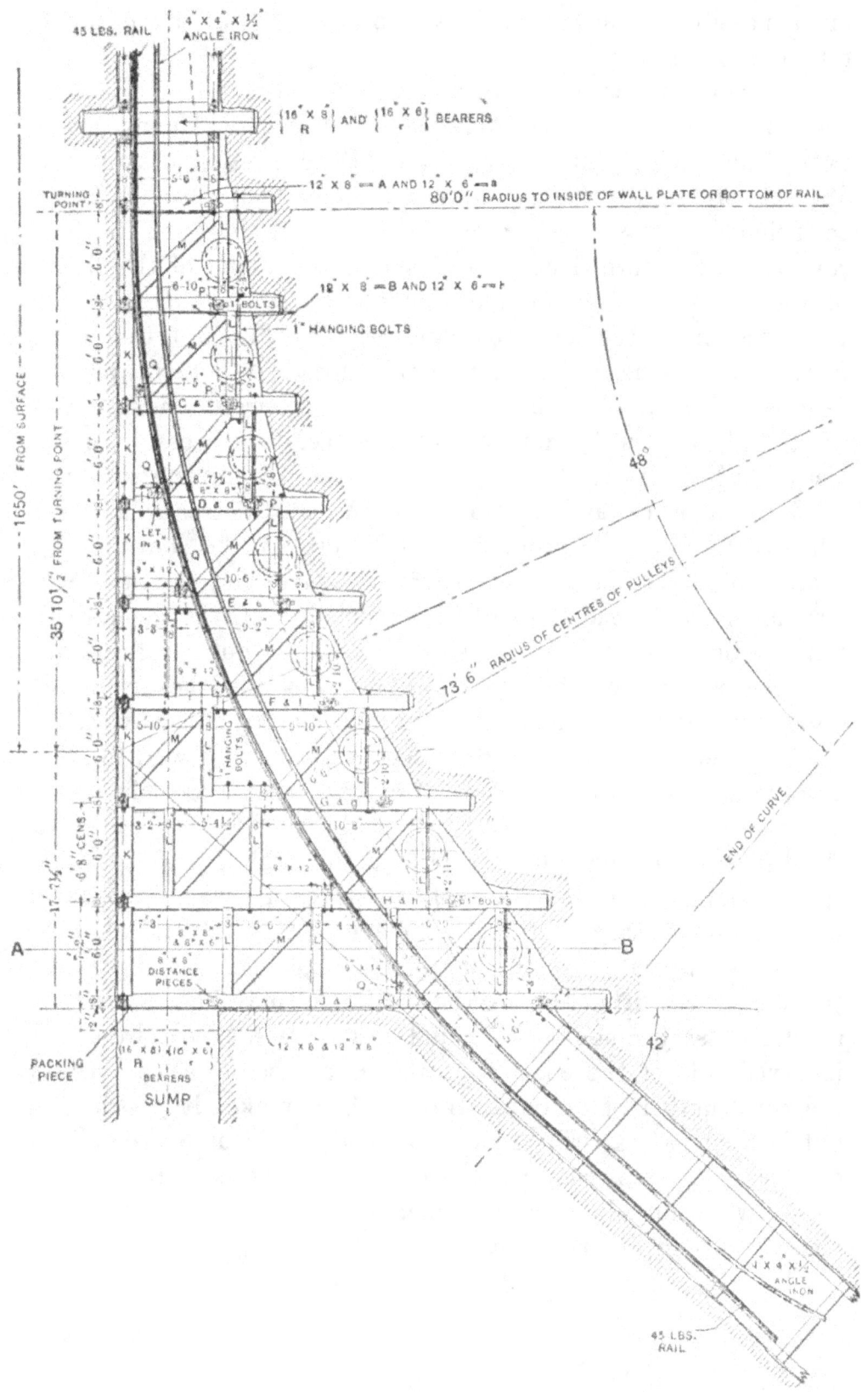

Fig. 50

far as possible the shock incident to a change of direction in the rapidly-moving skip.

This subject has been well treated by Thomas H. Leggett in the "Transactions of the American Institute of Mining Engineers," vol. XXX, under the title "Deep-level Shafts on the Witwatersrand," in which, concerning this change from vertical to inclined, he says: "Most of the deep-level shafts of the first row have been turned upon the incline, and many of the shafts of the second row will unquestionably be so turned; hence, I have shown in the accompanying sketch (Fig. 50) the usual method of making and timbering this turn. Its advantages are numerous. It avoids the expense of a separate underground hoist, and enables development on the reef to proceed with the least possible delay. The skip runs from the incline into the vertical smoothly, and generally without any decrease in speed whatever; hence it is probable that its use will be continued in shafts up to 3000 ft. vertical depth. After that depth, in all probability, independent methods of hoisting on the incline will have to be adopted. At the same time, it is not at all certain that the use of the Whiting hoist will not permit the turning of vertical shafts upon the incline, even at this or greater depth."

The sketch illustrates the manner of timbering work of this description. It may be varied, of course, to suit any change in conditions at various places. Particular attention is called to the disposition of the pulleys on the hanging-wall side of the shaft, the function of which is to carry the hoisting rope around the curve without unnecessary friction or damage to the timbers. It also shows the disposition of the angle-iron guide rails which prevent the skip wheels from leaving the track. It may be argued that an unnecessary amount of ground has been removed from the foot-wall side of the curve, but this is advisable, as it permits the construction of a timber frame which makes it possible to put in a curve that will remain absolutely rigid under the stress of work, with skips running at a high rate of speed. If the timbering were not placed in this manner a much lower degree of security would be possible.

Chapter X

BEARERS IN SHAFTS

It must be understood that, while blocking and wedges are employed to hold shaft timbers firmly in position, that any alternation of wet and dry will have the effect, during the dry periods, of causing the timbers, blocks and wedges to shrink, sometimes to such an extent that the latter become loose and drop out; security is then at an end, for the timber sets then have a tendency to settle by gravity, and every little subsidence or disarrangement is increased from set to set until there is danger of collapse of the entire structure. To obviate this, what are known as "bearers" are introduced, usually at each 50 ft. Where stations are 100 ft. apart, a bearer half way between is considered sufficient to support the sets dependent upon it.

The bearers are heavy timbers extending across the ends of the shaft, resting their ends in hitches cut in the solid rock. The bearers must be perfectly level and are so placed that the ends have a solid bearing on the hitch of at least 6 in.; more is better. The more loose the rock in which the sinking is being done, the greater the necessity for firmly placed bearers. Upon the bearers the ends of the wall plates rest, the end plates being placed in the usual manner. No change need be made in the framing of the wall plates, as the shallow dap cut on the under side to receive the post now rests in a similar dap cut in the bearer to receive the wall plate. These daps keep the wall plate from shifting in either direction. On the under side of the bearer daps are cut to receive the posts of the set beneath. The bearers should be at least as heavy as the plates, and greater depth is advisable, as the timber is weakened a little by the daps cut on both upper and lower sides.

Where shaft plates are 12 × 14 in., placed with the broad side up, the bearers should be 14 × 16 in., placed with the 14-in. face up. This makes the upper surface conform to the size of the shaft timbers and affords the additional strength which may be necessary when the hanging-bolts have been removed, should any shrinkage take place in blocking or wedges, or should the support of the shaft walls be weakened from any other cause. The bearer must be strong enough to support the weight of the ten sets of timber resting upon it — 50 ft. of the usual shaft timbering. This is the theoretical requirement, but it is seldom reached, for under almost any circumstances that can be conceived, it is difficult to imagine all support removed from each and every one of the ten sets. However, it is well to put in strong bearers, and to remember that their usefulness wholly depends upon their being set upon solid rock and firmly wedged. Shaft construction is far more careful and exacting work even than that usually done in the erection of frame buildings. Shafts are likely to be subjected to strains and tension such as no building is ever required to sustain, and this demands that the timbering be properly framed and placed with precision.

There is one more important matter in relation to sinking shafts through ground which at the time gives little unusual trouble, but which, it is known from its character, will cause no end of trouble later unless a method of timbering be adopted that will make it possible to meet the conditions that are expected to develop. Reference is here made to swelling ground, which, miners generally will agree, it is impossible to hold with timber, except where provision is made in the system of timbering that will make it possible to relieve the ground as it gradually, but irresistibly, encroaches upon the shaft.

There are different kinds of ground that swell, among them rotten granite (rarely), clay, foliated-slate, slaty-gouge, serpentine, and slate itself. Some of these swell more speedily than others upon exposure to the atmosphere, but in the opinion of the writer none of the various kinds ultimately give more trouble than the slate, which, firm enough when first cut, within a week or two shows signs of crowding the timbers, and within a month begins to bend, displace and to break them.

On the Mother Lode of California, particularly in the counties of Amador and Calaveras, this kind of ground is well known, and the experience of more than half a century in that district should have developed methods of dealing with the difficulty which would be of value elsewhere. The shafts on the Mother Lode which experience this difficulty with swelling ground are mostly inclines sunk in the vein fissure — sometimes in ore, quite as often in gouge. This trouble in the vertical shafts is usually only encountered where the shaft penetrates the vein, as it is adjacent to the fissures (whether filled with ore or only with barren foliated slate or gouge) that the rock swells.

The old Wildman shaft at Sutter Creek and the Keystone main shaft, at Amador city, afford excellent examples of the difficulties and expense attending the maintenance of shafts under the conditions met here. Both of these famous old shafts were originally timbered with heavy timbers, but all of the original timbers have been removed and replaced by others many times. The size of the timbers was also increased from time to time, until these shafts to-day are practically cribbed for hundreds of feet with great logs 24 to 30 in. in diameter, and even these fail to hold the ground, but must be removed from time to time, the ground relieved by cutting it away, and the timbers replaced. The heavy expense of thus constantly retimbering these old shafts is enormous. It was estimated that the necessity of operating through the old Wildman incline cost the company annually between $15,000 and $20,000 more than would have been necessary had the mine been equipped with a suitable vertical shaft, and it was this fact that caused the Wildman Company to commence sinking a large three-compartment vertical shaft.

We have pointed out the danger and expense attached to driving and sinking through swelling ground. It is true that usually both the danger and expense come some time after the cutting has been made — from a few hours to several weeks, but when the pressure begins to assert itself, it is irresistible, and proper means to deal with ground of this kind must be provided, or serious consequences will inevitably result.

It was learned long since, in drifts that had been driven through swelling ground, that timbers closely placed, with close lagging, could not be depended upon to hold ground of this character; that the more substantially the drift was timbered, and the closer it was lagged, the more certain was it that the pressure upon the sets would first cause the lagging to bend, and later, if they were not taken out, one at a time, and the ground cut away with a pick, the heavy timbers would be displaced,

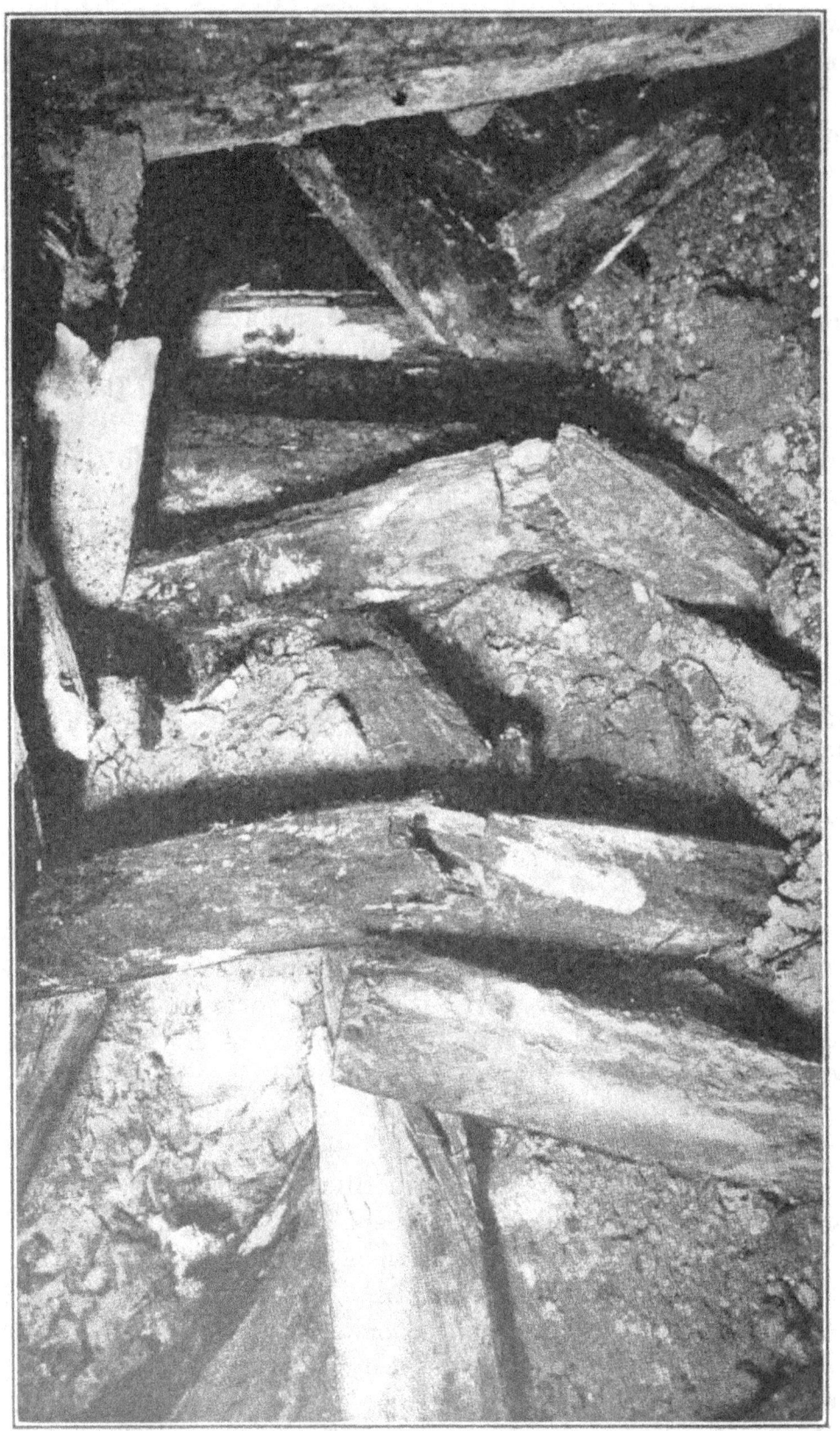

Fig. 51. — Drift Timbers Crushed by Pressure

bent, split or broken. This is illustrated in Fig. 51. It took years to learn that this trouble and expense could be almost entirely obviated if the lagging were put in with a space of 4 to 6 in. between; that through these open spaces the encroaching ground would slowly force itself, and that it might readily be cut away without removing any of the lagging, and the pressure upon the sets relieved, when necessary, at minimum expense. The principles employed in timbering drifts in this manner have been adapted to shafts. It has been learned that a shaft may be sunk through swelling ground, and that it may be kept in almost perfect alinement, if only the necessary precautions be taken when the shaft is being sunk. This is accomplished by cutting

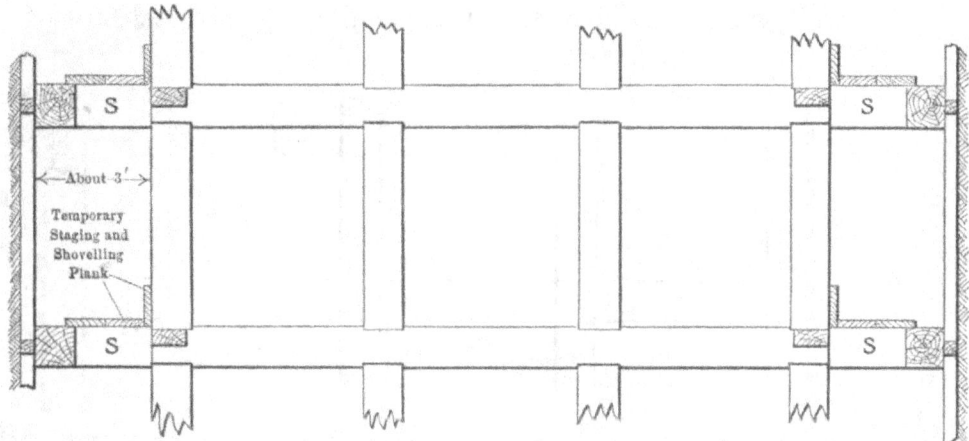

Fig. 52. — Longitudinal Section

the opening large enough to permit the building of a "crib" around the shaft sets, with suitable arrangements for relieving the ground as the pressure asserts itself.

The accompanying sketches, Figs. 52 and 53, illustrate a portion of a shaft that is protected by a crib built outside the main sets, Fig. 52 showing a vertical section of one side. The timbers of the regular shaft set are seen reinforced, as it were, by a second end plate which is separated from the end plate of the set by a sprag (S) about 24 in. long. The outer plate must be blocked and wedged where necessary, in the same manner as is usual with shaft sets. The blocking and wedges are omitted from the sketches purposely, as these would be placed where required. Fig. 53 shows a plan of a portion of the shaft. The lagging is placed outside the outer plates. It should be at least 4 in. thick and 6 to 8 in. wide. They should not be placed nearer than 6 to 8 in. from each other. Upon the sprags — one being placed directly opposite each end plate, opposite each end of the wall plates and also opposite the dividers — a platform of 2-in. plank is laid, one plank nearest the shaft being set upon edge to prevent rock or earth from falling down the shaft at any time. Whenever necessary men go upon these platforms and cut away the ground that is pressing against the lagging and forcing its way between them. The lagging may be removed and replaced after cutting out the ground. This method of timbering makes the work of relief comparatively easy and safe. As the ground is removed it falls upon the platform, from which it can readily be shoveled into a skip in the shaft. It will be conducive to safety to allow the platforms to remain in place — at least on every

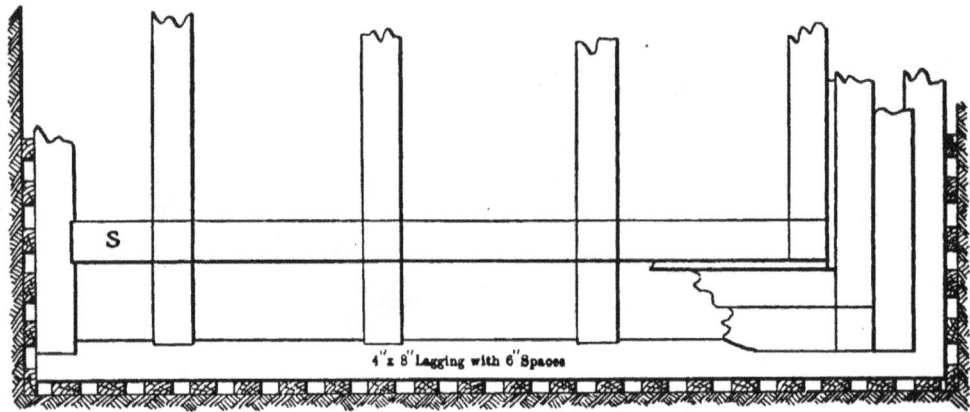

Fig. 53. — Transverse Section

other set — for the purpose of catching any rocks or material coming through the spaces between the lagging, preventing it from falling down the shaft. Of course, close lagging may be spiked to the main shaft sets, if considered necessary, but if the shaft be well timbered, in the manner suggested, in some instances, at least, inside lagging will be unnecessary.

Absolute safety to men in the shaft is the first essential, the consideration of economy being next, convenience in caring for the shaft being the last consideration; but to be economical the arrangement must be convenient. In a shaft timbered as here suggested, it will be advisable to use permanent hanging-bolts, as the ground in course of time is likely to exert unequal pressure and induce movement of the shaft sets, and this tendency can be much more readily controlled if the timbers are solidly suspended from bolts. It is probable that in a deep shaft a crew of two or more men may be required all the time to keep the shaft properly lined up. There are mines on the Mother Lode of California where from four to six men were employed constantly on shaft relief and repairs, working under the old system of heavy timbering to resist pressure. Whenever "cribs" have been introduced, they have effected a change for the better that was at once noticeable.

In inclined shafts, quite frequently, the pressure is from one side only — one side being hard rock. The old Eureka mine at Sutter Creek, for instance, has a hard, firm hanging-wall of metadiabase, and a foot wall of soft, black slate that swells upon exposure. That mine has been idle for about 30 years, and it would be no surprise, should it ever be reopened, to find that practically all the workings in the main fissure have been completely closed by the swelling slate. Where the swelling ground is on but one side, then the crib will only be required on the disturbed side. It has been suggested that in such a fissure it would be a good idea to sink a shaft with the longer axis of the shaft normal to the fissure, that is, the end to be in line with the direction of pressure. In a vertical shaft this scheme might be introduced to advantage, but in an incline it necessarily contemplates one compartment overlying another, with the pump and manway compartment on the bottom, and the hoisting compartments one above another, overlying the pump compartment. The theory advanced is, that as a much smaller area would be directly exposed to the region of pressure, it would be proportionally easier to hold the ground in a shaft of relatively smaller sectional area. Just how this would work out is a question, but it seems that the pressure, which in all probability would be exerted also from the direction of the strike of the disturbing formation, would be likely to prove a source of much trouble.

Fenders for the Protection of Shaft Timbers when Blasting

Shaft sinking always requires that some precautions be taken to protect the timbers at the time of blasting, where timbers are carried within 20 ft. of the bottom or closer. This may be done in various ways, the most common being the hanging of green pine timbers on chains beneath the plates of the lowest set; or the fender timbers may be suspended by means of bolts passed through holes bored through the fenders and run upward through the unused hanging-bolt holes of the lowest plates of the sets above. The bolts should be L-shaped, having a thread cut at the long end only, so that a nut provided with a handle may be screwed upon it to hold the timber in place. The lower end turned outward at a right angle will support the timber.

A third, and very satisfactory way is to place steel plates beneath the last set and also on the inner face of the timbers. These steel plates should be in sections and bent at a right angle, so that they will protect both bottom and sides from flying rocks. Each of the methods above mentioned affords some protection to timbers of the lower set, which is more or less satisfactory, according to the care used in placing the fenders, but neither of these schemes affords the slightest protection to sets above the lowest, whereas it is of almost equal importance to give some protection to the timbers of sets above for a distance of 25 or 30 ft. We have known of a wall plate in a set 25 ft. above the bottom of the shaft, and in the third set from the bottom, being broken squarely off by a flying rock. The ground was very hard and No. 1 dynamite was used. Under these conditions rocks fly with terrific force, and timbers must be either very large — unnecessarily so for the ground — or be well protected to withstand the bombardment of rocks when blasting.

The Cable Mat

In view of these conditions, it is desirable that some sort of protection be provided for all the timbers near the bottom of the shaft. This has been successfully accomplished by suspending a mat made by weaving together sections of old hoisting rope. This can easily be done with rope up to 1-in. diameter. A mat made in this manner can be drawn up by chains or pieces of cable and made fast, affording the best possible protection to the timbers. In the case of a large shaft, say one of three compartments, and measuring 16 ft. in length and 7 ft. in width, outside measurements, the mat might be made in two sections. In a single section, a mat of this size, 16 × 7 ft., would require about 250 ft. of cable; if made of 1-in. rope this would weigh about 400 lb. A similar mat made of $\frac{7}{8}$-in. rope would weigh 300 lb., and one of $\frac{3}{4}$-in. rope would weigh but 220 lb. If in two sections, mats of rope of these several diameters would weigh approximately but one-half of the weights above given. Mats in sections are advised for convenience in handling, not only on account of the weight, but for the reason that large mats cannot readily be taken up and down the shaft. They should be divided the long way of the shaft — each section made about 3½ ft. wide by 17 ft. long. These, hung side by side, will protect the timbers in a shaft 7 × 16 ft. outside of the timbers.

The mats should be suspended in such a manner that they are not likely to be unhooked while shots are discharging, allowing the mat to fall and leave the timbers exposed to the destructive effect of the blasts. To suspend the mats, either chains or pieces of wire cable of ⅝-in. diameter are heavy enough. These should be attached to the mat by rings at the corners and at or near the middle of the sides, these latter preventing the mats from sagging too much at the center. The upper ends of rope or chain should be provided with hooks which may be caught in the hanging-bolts of the second set from the bottom. The mats should not be drawn too close to the timbers, as greater protection is afforded if they hang a foot or more below them. The mesh of the mats should not exceed 9 to 10 in., and if somewhat closer, so much the better. Where these ingenious fenders have been used they have been found to afford all the protection to timber required, and their use is strongly recommended.

Extension Tracks for Sinking

The employment of fenders for the protection of shaft timbers has been fully described, and their usefulness explained in detail. There are other things in shaft sinking, which, small in themselves, help to make up the whole. It is by a thorough understanding of the details of mining work that we are enabled to accomplish the most work in a given time and thus reduce the cost of the operation, whatever it may be. In sinking vertical shafts, the removal of broken rock is best accomplished by means of a bucket guided by a cross-head, the method of making and the use of which have been previously explained and illustrated. In inclines, sinking is usually carried on by means of a small skip, though often a bucket is used which slides between two skids. When the last set of timbers is reached there must be some means of continuing the skip or bucket to the bottom of the shaft, which is done either by movable skids or by movable rails. This practice has led to the introduction of extension skids for buckets. These are generally made of wood and are often so heavy that the united efforts of several men are required to handle them, the greatest trouble arising when they have to be hauled up to a place of safety at the time of blasting. Not only are these extension skids made of heavy timber (generally 4 × 6 in.), but they are also often seen shod with heavy iron straps, and the two timbers are bound together by means of several heavy bolts. As a matter of course, the skids should be strong enough to carry the loaded bucket, but generally this part of a shaft-sinking outfit is made unnecessarily heavy.

On the Mother Lode of California, shafts sunk in the fissure are usually provided with wooden tracks faced with strap-iron. The object of this is to provide a run-way for skips which will be more enduring than the bare wooden rails. It is unsafe to use T-rails in these shafts, owing to the fact that the ground is constantly shifting, and this has been found to render tracks of T-rail very dangerous, as the projecting end of a rail may at any point in the shaft derail a passing skip.

In inclined shafts, sunk in solid ground, we believe that skips are preferable to buckets for sinking, and T-rails are always to be preferred to strap-iron tracks, except in such ground as that referred to as occurring on the Mother Lode.

Inverted Rails

In sinking inclined shafts in firm ground we have found that a very convenient extension track may be made by inverting T-rails and clamping them to the rails of the track above (see Fig. 54). At the upper end these inverted rails should have the base of the rail forged flat and fitted by the blacksmith, so that it offers no considerable obstruction to the passage of the skip wheels. These rails should be at least 20 ft. in length, and they should be kept from spreading by means of clamp rods similar to those used on railway tracks at switches, which may be easily adjusted whenever needed. Any unevenness in the bottom of the shaft does not affect the usefulness of these extension-track rails, as they are independent of each other, and one may move downward several feet further than the other; so that, no matter how uneven the bottom of the shaft is, the track may be kept perfectly straight and at the same angle as the main track. It is advisable to have several different lengths of track when sinking — a 5-ft. length, a 10-ft. length and a 15-ft. length. These are used in turn, a longer replacing a shorter length, as sinking proceeds; a 20-ft. length being placed permanently when sufficient headway has been made.

Sinking-Ladders

Another thing that requires careful and constant attention is sinking-ladders, and this applies to both vertical and inclined shafts. Sinking-ladders should be made of strong, light wood, and iron steps, made of pipe, are advisable, rather than those of wood, which break easily and are not at all adapted to the

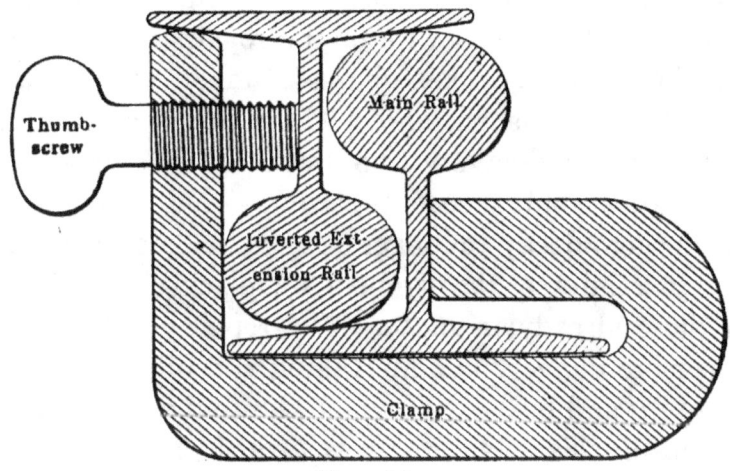

Fig. 54

construction of sinking-ladders, always subjected to more rough usage than the ordinary ladders of the mine. What the men require is a strong light ladder — one that is durable and easily handled. It is not necessary to make these ladders more than 12 in. wide between the sides, though 14 or 16 in. is usual for the permanent ladders.

Inclines in Hard Rock

Thus far we have said nothing of those inclined shafts sunk in hard rock in which only foot-wall timbers are employed, their chief and in fact their only function being to afford the necessary support for carrying the tracks, pipes, ladders and any other equipment placed in the shaft. Shafts of this description are not numerous; still there are such. It is merely the good fortune of the mine owners that they need no more timbers in the shaft than those used as foot-wall plates. It is usual to set up a line of props or stulls between the compartments of a shaft of this description, the object being to afford a line of supports for the signal-bell cord, but principally to lessen the always present danger to men passing through the shaft, which would be greatly increased by the absence of the line of props.

Shaft Repairing

Shaft repairing is often more difficult than shaft sinking, but experienced men can generally overcome any difficulty likely to be encountered. It requires generally the employment of temporary supports to hold the ground while the old timbers are being removed to make room for the new ones. Of course, it is important what the nature of the repairs is. If it is merely the removal of a single broken or weak timber, the job can usually be promptly accomplished, and if the time be well chosen often without materially delaying the work in the mine. At other times the entire shaft needs repairing — old timbers becoming too weak through decay to longer be safe. Sometimes it is merely necessary to reinforce the timbers by placing intermediate sets between those already in place. The repairing of a caved shaft is a much more dangerous undertaking and often calls for all the skill, ingenuity and bravery of which the timber crew is possessed. An instance of this character was furnished a few months ago at the Alpha shaft of the Giroux Consolidated mines at Ely, Nevada, where several hundred feet of the shaft was lost by caving, but which was recovered by the prompt and heroic work of chosen men who labored unceasingly to rescue those imprisoned below. The history of mining has furnished many such instances of the skill and heroism of miners. They are never found wanting when human life is at stake.

Chapter XI

POSITION AND DIRECTION OF DRILL HOLES IN SHAFT SINKING

The disposition of drill holes and their direction in sinking shafts is of importance. In handwork this is done with particular reference to each part of the shaft, and the holes will vary more

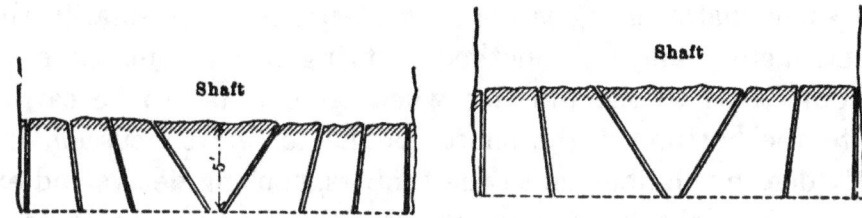

Fig. 55. — Showing Direction of Drill Holes

or less with each round, both as to the number of holes drilled and their direction and depth. But in machine drilling it is

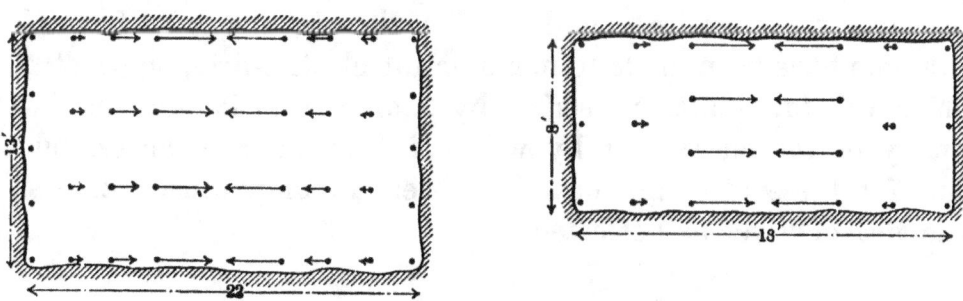

Fig. 56. — Disposition of Drill Holes

customary to drill the full round before blasting, whether the entire round be fired at once or not. When machines are employed, the holes should be arranged systematically, in order to secure the best results with the least expenditure of time and money. Figs. 55 and 56 illustrate the usual manner of placing holes. Two separate shafts are shown — one 18 × 8 ft., the other 22 × 13 ft., outside of timbers. Obviously, more holes are generally required to break the rock in a shaft of large section than in a small one. The holes nearest the center (the "cut") are fired first, then the others, either in series or simultaneously.

Many engineers believe that shaft sinking can be accomplished more rapidly and at less expense by drilling the "cut" holes near one end of the shaft, and firing the blasts in series from the cut across the cutting toward the opposite end. This usually results in throwing the muck pretty well to one end of the shaft, so that by mucking at the end where there is the least debris, space is soon made to set up machines and continue drilling while shovelers clean up the opposite end. This was done at the Hoatson shaft of the Superior and Pittsburg Company, at Bisbee, Arizona, and a record for shaft sinking established for that district in the Hoatson shaft.

In some shafts, firing is done by electricity. Ordinarily this is a satisfactory and safe method of firing blasts, but there are some objections to the practice when timber has to be carried close to the bottom of the shaft, as the terrific concussion frequently does much damage to the timbers, causing delays and expense which might have been obviated by cutting the fuses to suitable lengths — the short ones in the cut holes and longer ones nearer the ends. Where the holes are fired in series, electricity may be safely used, and in shafts where the timbers are not too close to the bottom, there is an advantage in simultaneously discharging the entire round by the electric method. Several attempts have been made to manufacture electric-firing apparatus which will fire blasts in series, by a differentiation of the intensity of the current or by wires of differing resistances, but thus far these attempts have not been an unqualified success. The idea may yet be perfected.

Chapter XII

CUTTING AND TIMBERING STATIONS AT SHAFTS

Not the least important matter in the sinking and equipment of shafts is the cutting and arrangement of stations. In recent years it has come to be recognized that much in the economy of mine operation depends upon the arrangement of the station. It is not so very long since it was a common thing to see stations at the shafts of important mines merely an enlargement of the drift, affording a safe passage for the workmen to pass the shaft in going from one side to the other. Often the only arrangement for handling ore consisted of a small, sloping, open chute, with an extension of a few inches into the shaft, this projection or apron being sometimes fixed, and in other cases provided with hinges so that it may be turned back or dropped down, out of the way of the skip or bucket. When it was desired to load the skip or bucket with ore, it was stopped at the level and a car of ore was dumped upon the chute, from which it ran or was scraped into the waiting skip or bucket. This was a crude and unsatisfactory method of handling ore, and it is evident that it was impossible to handle a large tonnage in this manner. As a matter of fact, comparatively few mines were equipped in this fashion; still they were numerous enough, and it was one of the recognized methods of ore handling at shafts.

Cages, upon which cars are run, have been in use for many years, having first been introduced in Europe, where the cages are generally ponderous affairs. In America the cage is made of steel, is light and durable, and affords a very satisfactory means of handling ore. Cages are made with one, two, three, and even four decks, so that several cars may be run in turn upon the platforms of the cage, and hoisted at one time. When cages were first introduced, about 1860, they were at once recognized as so superior to the old shoveling and flat-chute arrangements that there was "no comparison" between cages and former methods. In referring to the introduction of cages in mines on the Comstock Lode, Dr. R. W. Raymond said, in his report to Congress, 1870: "The mineral, having been taken to the shaft, is either dumped in a pile and then shoveled into the bucket or skip, or is dumped through a chute directly into the skip, and the empty car is returned to the face. But this necessitates a rehandling of the mineral, which, when it reaches the surface, must be dumped again into a car or wagon, by which it can be delivered at the proper point away from the shaft. These, and other considerations, have led to hoisting the car and load together to the mouth of the shaft. This effects a great saving of time and labor, and wear and tear of apparatus. It is the method adopted in the mines upon the Comstock Lode, and in all well-appointed vertical shafts of any considerable depth elsewhere."

Ore Pockets Beneath the Levels

This was before the day of cutting ore pockets beneath the stations, and before the day of automatically loading and dumping skips. It is true that skips had, even prior to that time (1870), been arranged to dump automatically upon reaching the surface, but the skips were comparatively small, and the running speed of the hoisting engines was not great.

To-day the ore pocket beneath the stations, skips of large capacity, fast-winding engines, and automatic devices, have come to be recognized as necessary to the economical hoisting and handling of ore. So important has this become, and so great the decrease of cost from that of hoisting by cages, that many of the largest and deepest mines of the United States have within a recent period made expensive changes in their shafts, and have substituted skips for cages. Among the mines having made or making this change are several at Butte, Montana, the Calumet and Hecla copper mine at Houghton, Michigan, with its mile-deep shafts; the Copper Queen mine, at Bisbee, Arizona, and the Homestake, at Lead, South Dakota. It is not a matter of preference, merely, but, it has been clearly and repeatedly demonstrated that hoisting a given tonnage with skips can be done more rapidly and at less expense than by any other means at present in use.

Stations should be cut at the time of sinking. When the shaft reaches the point where it has been decided to establish a level, the ground should be broken out a few feet on that side where it has been planned to cut the station. Working room only need be provided, which is at the time sufficient. The work of sinking should be continued until the shaft has been completed to a depth of at least 30 ft. below the floor of the level it is intended to establish. In the fifth or sixth set below the level, strong chutes should be built. It is well to provide these with steel doors, operating with rack and pinion. When all is completed, and there is sufficient room to accommodate the skip beneath the lip of the chute, the miners may go into the recess back of the timbers, and for protection line up the shaft sets with heavy green logs, or old T-rails, and proceed to cut out the sloping floor of the ore pocket. This should have an angle of at least 45 degrees, so that if the bottom of the chute is 25 ft. below the level, the inclined bottom of the pocket will extend backward to the door of the level at least 25 ft. from the shaft. The miner should at first do all drilling by hand until sufficient room has been made to employ machines, by which latter means two-thirds of the work, at least, may be done. It will always be necessary to protect the shaft timbers as far as possible from injury by blasting, and for this reason holes must be pointed and loaded with good judgment, avoiding the use of an excess of powder. The writer has had stations cut in shafts that had been sunk several hundred feet below the level of the station, and little injury to the shaft sets resulted, where precautions were taken to avoid it, so that we know work of this kind can easily be done by experienced miners.

The character of the ground to be removed may have an important influence on the methods employed and on the speed of the work, for it may be necessary to put in many temporary timber supports to hold ground until the main timbers can be placed. It will be evident that all rock broken in the station from the bottom of the ore pocket to the roof of the station may be drawn out through the chute doors, thus dispensing with much shoveling and consequently reducing the cost of cutting the station. Even of the rock blasted in cutting the station itself, back of the top of the inclined bottom of the ore pocket, the greater part will fall into the pocket and may be drawn into the skips without shoveling a pound of rock. The method applies to either inclined or vertical shafts.

Stations should be roomy—there is little economy in cutting them too small. The station should be at least as wide as the shaft, and, on the side adjacent to the hoisting compartment, should extend fully two feet beyond the shaft. This affords a safe landing place for men, who may step from the skip upon the platform at the side of the shaft instead of directly in front, where there is sometimes danger of falling into the ore pockets beneath. The space at the side also affords convenient room in handling materials going into or coming out of the skips, and this applies to inclined shafts as well as to those that are vertical.

In height the station should be not less than 12 ft. at the shaft. This is absolutely necessary in stations at vertical shafts, and even a greater height is often an advantage. In inclines, stations may be somewhat lower, but unless the shaft be very flat — less than 40° — it is a mistake to cut the station less than 10 ft. high adjacent to the shaft. This permits the unloading of long lengths of pipe, rails and long timbers, for which there must be sufficient head room. Every shaft station should be equipped with a strong block and tackle, for the convenient removal of heavy articles from the skip. The block and tackle is suspended from the roof of the station, properly from a stout hook, one opposite each hoisting compartment. Where machinery is to be handled, chain blocks should be provided. These little conveniences greatly facilitate and cheapen labor, and failure to place them in stations is mistaken economy.

From 10 to 12 ft. back from the shaft the roof of the station may be somewhat lower, though it seems to be a mistake to ever cut the opening less than 9 ft. clear of the track. A low back in a station has many disadvantages. It directly affects ventilation, and is often regrettable, particularly where long pipes and timbers are to be handled. In width the station should be, as previously stated, as wide as the shaft, and the width should be maintained well back of the ore pockets. At the rear end it is a good idea to have a stout work bench, provided with a vise and other tools, and a small stock of pipe fittings may be carried in boxes arranged above the work bench. This is to facilitate the making of small repairs to pipe lines or to ventilating apparatus. The repairing of machine drills should be done in the machine shop on the surface, unless the mine be provided with a fully equipped shop and forge underground — a practice that in recent

CUTTING AND TIMBERING STATIONS AT SHAFTS

years has been quite generally adopted in many large mines. In the Homestake mine at Lead, South Dakota, there are two large and well-equipped shops underground, one at the 500 level and one at 800. In these shops all ordinary repairs are made, hundreds of drills are sharpened daily by mechanical drill sharpeners, and much work done which would otherwise go to the shops on

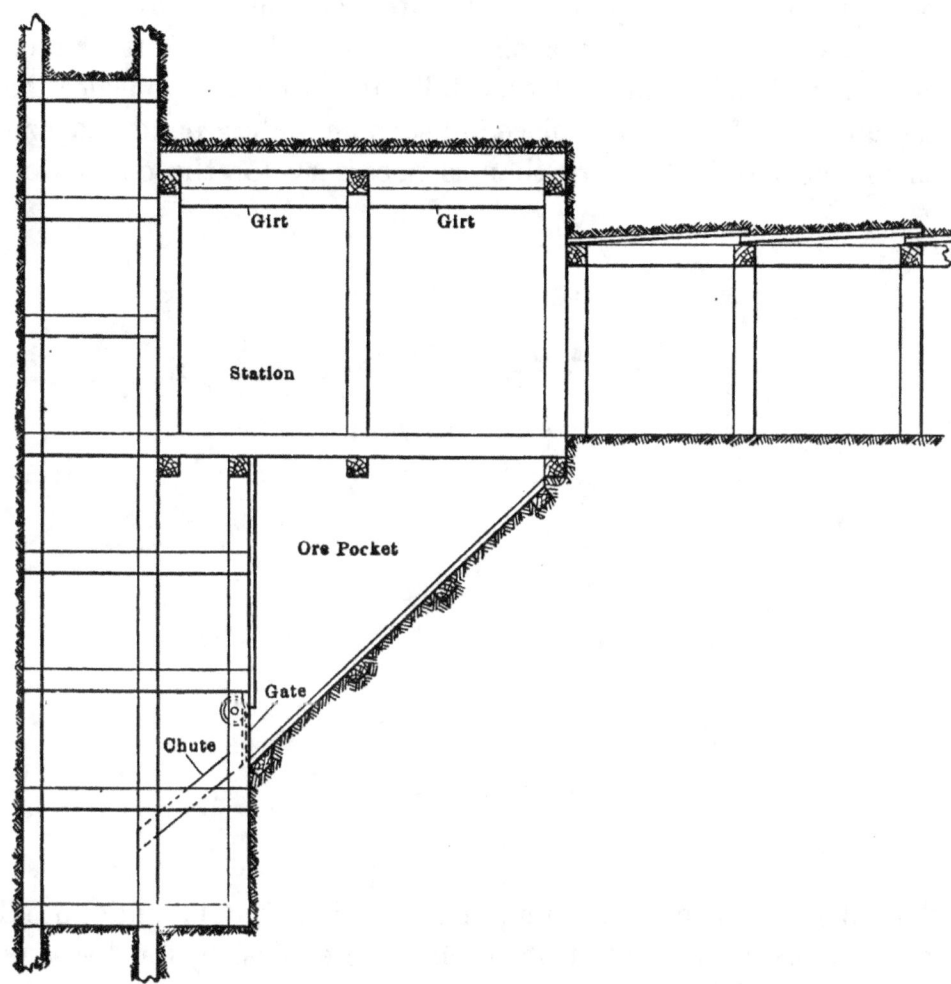

Fig. 57

the surface. The saving in time and otherwise amounts to thousands of dollars annually.

In small mines it is customary, whether it be good practice or not, to cut fuses and fit them with caps at the work bench in the station, and in numerous instances the criminal practice of keeping several boxes of powder in the station is permitted. This latter should be prohibited by law, as many fatal accidents are directly due to the careless handling of powder at stations.

Powder and caps should always be kept in separate places, and while it is true the powder can easily be exploded without the presence of the caps, the danger is greatly increased by having the two in close proximity. It is the best way to have a small room in a dry place in each level where the fuse and caps may be prepared by a man whose duty it is to attend to this, and the powder should be kept in a separate place and away from the main lines of traffic of the mine — an abandoned drift or crosscut, or a chamber cut out especially for storage of powder, and the amount of explosive placed in such an underground magazine should be but little more than is necessary for the day's work.

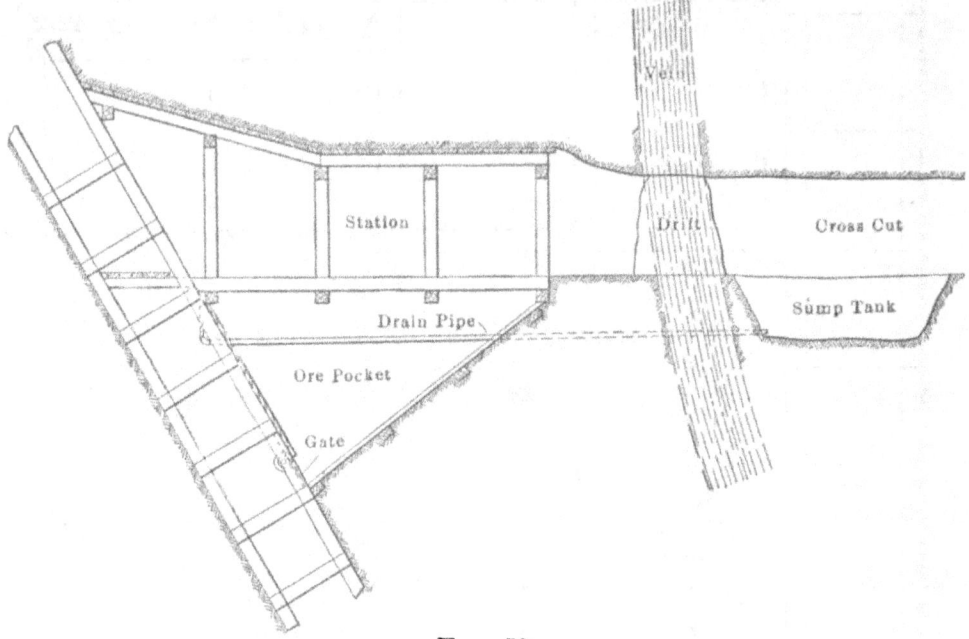

Fig. 58

The observance of these simple suggestions would undoubtedly prevent many accidents from premature explosion, for disastrous results have followed in scores of mines where these precautions have been neglected.

Of course, stations cut in hard rock require comparatively little timber, but in some mines this fortunate condition does not exist, and the stations require more or less elaborate timbering. No rule need be laid down for the timbering of stations, as the timbermen will naturally employ such methods as the conditions require. The principles underlying the timbering of underground excavations will govern in the sustaining of stations as well as other cuttings.

Types of Stations at Inclined and Vertical Shafts

Having said all that it seems necessary to say about stations at both vertical and inclined shafts, we here illustrate the two types, each as representative of its class. In the sketch of the station at the vertical shaft, Fig. 57, it will be noticed that there is a considerable recess between the front of the ore pocket and the shaft proper. While it reduces materially the storage capacity of the pocket, it affords opportunity for prompt repairs about the timbering of the pocket, and also provides a safe place for the skip-tender while in the performance of his duties in signaling and loading skips. These two points are distinctly in favor of constructing loading-bins and chutes beneath stations after the manner here suggested.

The sketch of the station at an inclined shaft, Fig. 58, is a duplicate of those made under the direction of the writer in a mine in Amador County, California, and which were found to meet every requirement. The pocket was divided by a partition through the center so that when desirable waste could be dumped in one side and ore in the other. When no waste is being sent to the surfaces, both sides may be used for ore.

Draining by Means of Skips and Tanks

At the rear of the station, and beneath the level, is seen an excavation, cut as a sump to catch the water coming to the station from that level. A trench was cut through the rock connecting the sump with the ore pocket and a 6-in. pipe laid from the sump to the shaft. The pipe was given a very small grade, and at the end was secured a piece of canvas hose about 3 ft. long. A plank and clay dam made the end of the tank tight. A loop of rope was made fast to the end of the hose so that it might be hung up when not in use, the end then being higher than the valve at the further end of the pipe, though if desired it may be higher than the level of the water in the tank. The flow of water was controlled by a simple clack-valve placed at the end of the pipe and submerged in the sump. This was operated by a rope which passed upward to the roof of the level and thence along the station and down the shaft, where the skip-tender could manipulate it conveniently. When the skip arrived he simply unhooked the hose, dropped the end into the skip, and pulled the valve rope.

The water rushed through the pipe into the skip. When it was about full the rope was released, the valve closed, the hose after draining a little is hung up, and the skip rung away.

Mr. Hans C. Behr, in his excellent work on "Mine Drainage Pumps," Bulletin No. 9, of the State Mining Bureau of California, illustrates a well-arranged sump and its attachments for draining a mine in this manner. The sketch (Fig. 59) is from his book.

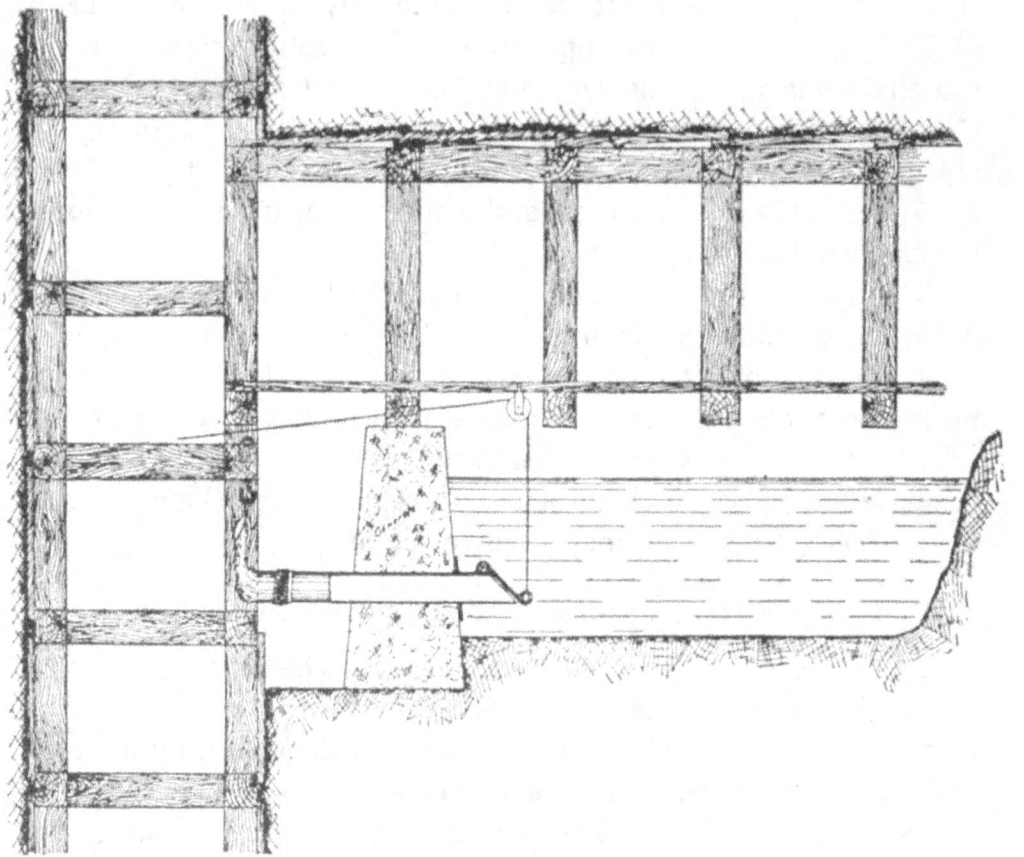

Fig. 59

It should be stated that when the pipe from the sump to the shaft passes through the ore pocket, it must be protected from rocks dumped into the pocket from the cars on the level above. This may be done by running the pipe either through the center of the pocket next to the partition, or at one side adjacent to the wall of the pocket. In either case it should be protected in such a way that it will be shielded from impact of the rocks.

While speaking of mine drainage it seems advisable to again call attention to the advantages of bailing. We have referred to this heretofore, but believe its importance justifies further mention of it. Mr. Behr, in the work above referred to, has devoted considerable space to the subject, and several of the sketches accompanying his remarks are reproduced. He says: "The simplest method of raising water from deep mines is by means of bailing tanks which may take the water either from station reservoirs or from the sump. In the latter case they are either made self-

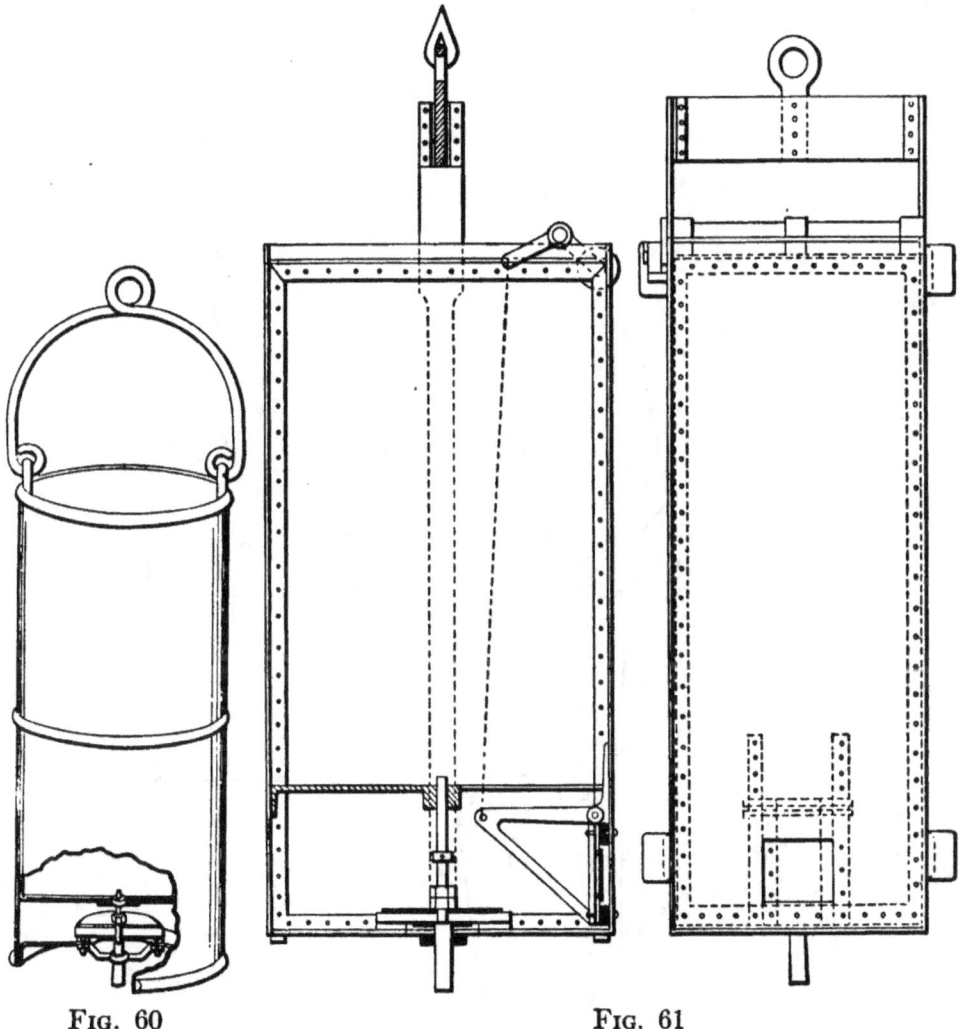

Fig. 60   Fig. 61

filling, or are filled by means of pumps or other contrivances. Rapid filling of bailing tanks is very important. Figs. 60 and 61 illustrate types of tanks fitted with a valve in the bottom, which opens of itself when the tank sinks in the sump water. Such a method requires a considerable depth of sump in order to fill them, and tanks are not, therefore, adapted for sinking unless artificial means be used to fill them.

"The discharge from bailing tanks at the surface or at a drain tunnel must be effected in such manner that none of the water will fall down the shaft again. The bucket in Fig. 60 has a downwardly projecting stem on the valve, which strikes the

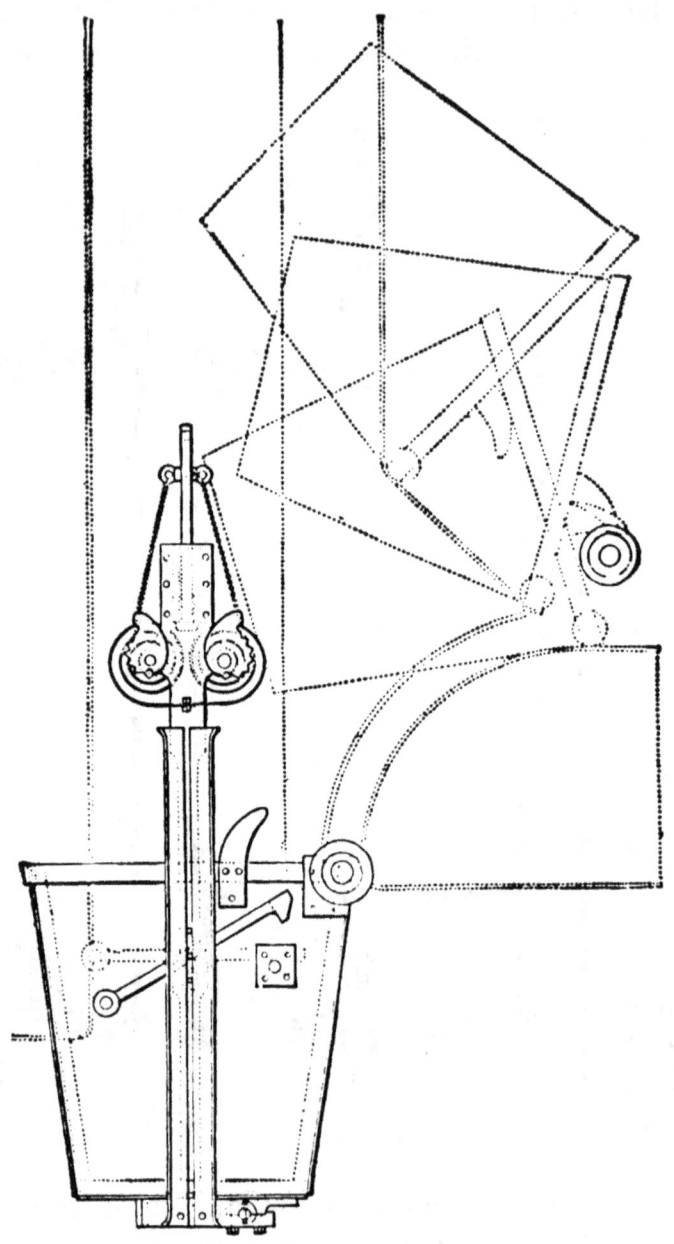

Fig. 62

floor and lifts the valve when the bucket is lowered into the discharge sluice, thus permitting the water to escape. Bailing tanks guided in vertical shafts usually have side valves, as in Fig. 61, and these sometimes have attached a hose which leads the water into the discharge sluice. The most rapid manner of discharging bailing tanks is to construct them like skips, so that they will dump automatically as they are hoisted above

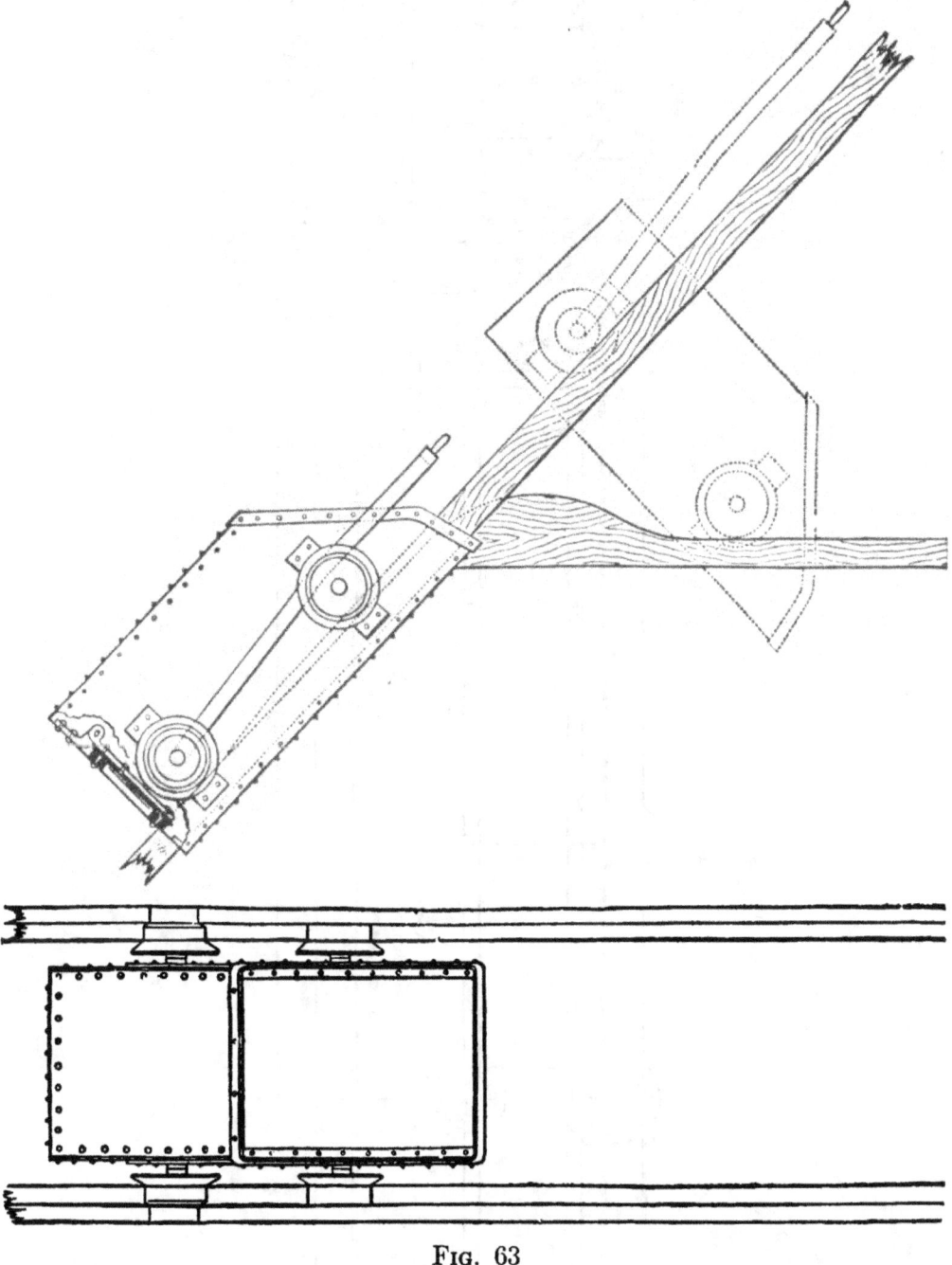

Fig. 63

the collar of the shaft. Fig. 62 shows a self-dumping skip for a vertical shaft which may be used for either ore or water, and Fig. 63 illustrates a simple skip used for inclines. The method of

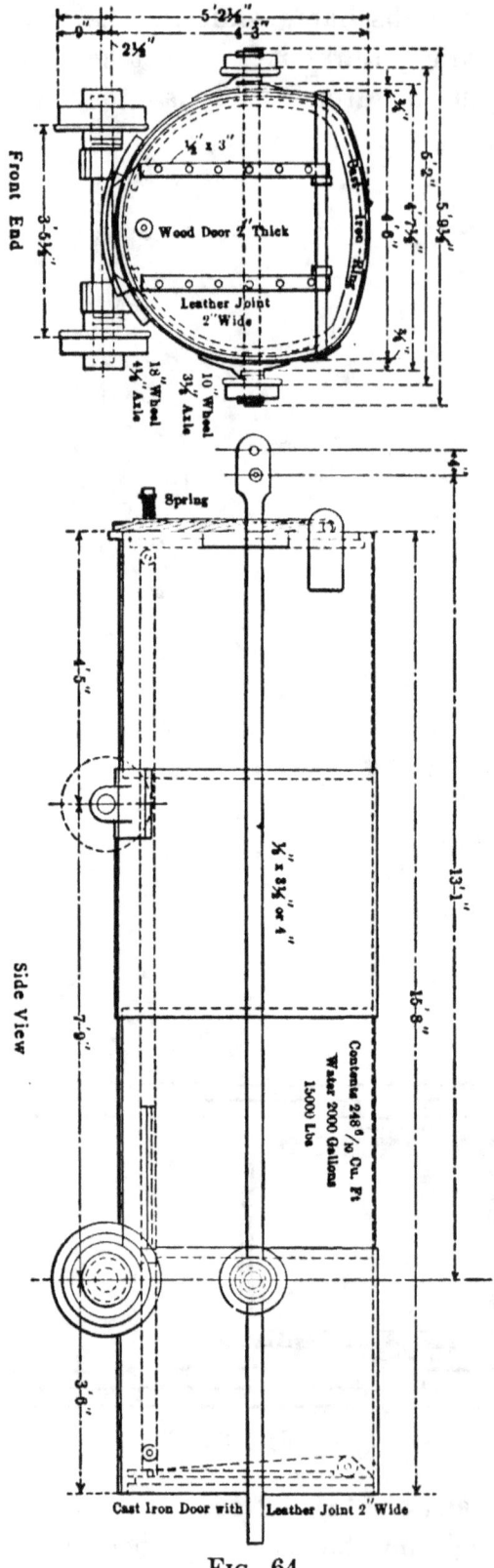

Fig. 64

dumping these skips is also shown." (Fig. 62 illustrates the Comstock skip, and Fig. 63 that in common use on the Mother Lode of California.)

Bailing by skips connected tandem is practised in some of the large Pennsylvania coal mines. Both the details of construction and methods of dumping are shown in Figs. 64 and 65.

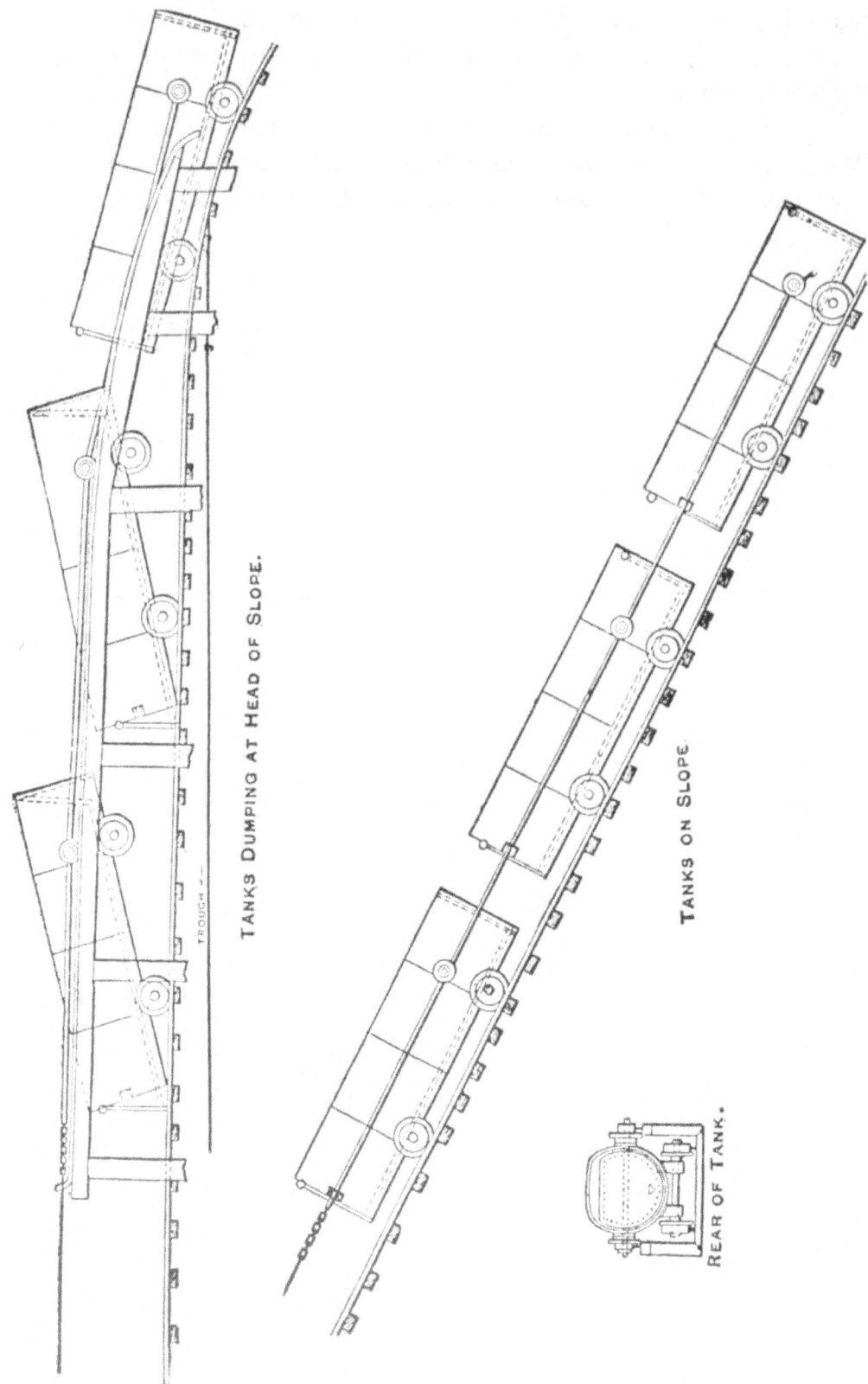

Fig. 65

Chapter XIII

MINING LARGE ORE BODIES BY THE OPEN-CUT OR "GLORY-HOLE" SYSTEM

The great majority of mines are opened on comparatively small veins of ore, and these are developed and operated for most part through adits — tunnels run from the surface on the vein — or across the formation to the vein, and drifts then turned on it; or the vein is attacked through a shaft, either vertical or inclined. Comparatively few small veins are worked by open pits, though there are such in numerous places. Occasionally, veins worked in this manner do not carry their value to depth, and in such instances deep workings are unnecessary.

Large veins and deposits outcropping at the surface are generally worked at and near the surface by open pits. The particular method selected in the mining and handling of ore in these large open excavations generally depends somewhat upon the topography of the ground immediately adjacent to the mine. It is the best practice, where possible, to drive a tunnel into the ore body and connect with a shaft sunk from the surface. If the tunnel level is the lowest in the mine at the time this work is done (it generally is) a chute should be built at the foot of the shaft or raise to facilitate the loading of cars. This chute should be strongly but not elaborately made. That is, no unnecessary expense should be incurred, as in most cases it is only a temporary arrangement, to be torn out as soon as the excavation of ore has reached the level of the chute or has passed the limit of convenience. The raise or shaft connecting with the tunnel should not be vertical. Many think that a vertical cutting of this description is the shortest and most direct way to make connections, and so it is; but while vertical shafts possess some advantage over inclines for the purpose of hoisting ore, they are not satisfactory as ore passes, for if a vertical chute becomes filled with ore, the chances of its jamming are excellent, and once in this unfortunate condition there are few miners, even the boldest, who relish the job of starting the rock again. It has been found by experience that an ore pass cut at 50 to 55 degrees is the most satisfactory, and while chutes at this angle also jam, they do so much less frequently than those that are vertical.

When the connection has been made from the tunnel to the surface, and the chute has been built, the mining of ore may proceed. The first thing to be done is to enlarge the shaft at the top by drilling holes and shooting them with small charges of powder, the object being to break the ore, but not to clog the chute by allowing too large pieces to fall into it. The process of enlarging this hole continues, and when it has been carried down several feet — twenty or more, a grizzly of logs should be put in to prevent large pieces going down into the mill hole. These are retained on the grizzly, where they may be broken up with hammers, or "bulldozed" to a size which will pass the grizzly. In many places the practice is to connect the tunnel below with the floor of a cut already excavated, placing the grizzly at the level of the cut. For a time it may be that the rock blasted down from the sides of the cut may fall or be shoveled directly upon the grizzly, the large pieces being broken up, but in time the working face will advance, and then either the grizzly must be extended or tracks must be laid to the working face, where the ore is shoveled into cars which are trammed to the chute and dumped. In the latter case no grizzly is required, as the large rocks are broken up before being shoveled into the cars.

Where the ore face is high, and a large amount of ore is available for a single chute, it is a good idea to have the raise come up at an angle from beneath the working face, instead of toward it. When arranged as here suggested, the ore may be sent down through the grizzly with a minimum of shoveling by simply extending the grizzly as the work advances, the ground overhanging the chute being cut out from time to time as required. At many mines grizzlies are dispensed with entirely, all the ore broken in the cut being shoveled into cars and sent to mill, either by dumping into a mill hole or tramming to a chute outside the cut. Often the breaker floor of the mill is on a level with the cut, so that the ore is trammed directly to the mill, being handled but once. Where waste that may be sorted out is mixed with the ore, it is better to shovel into cars, separating the waste by hand.

MINING LARGE ORE BODIES 135

Where this is necessary the tracks should be provided with several branches, an ore car to occupy one branch and a "stone-boat" the other. As shoveling proceeds, the ore is thrown into the car

FIG. 66. — Open Cut Big Indian Mine, near Helena, Montana

and the waste piled on the stone-boat for such disposition as may be desired. Where there are underground stopes in the mine, below the open-cut level, the waste should be dumped into a chute connecting with such stope, where it may be utilized as filling.

Among large mines employing the open-cut and mill-hole system are the Homestake, of South Dakota, the Big Indian, near Helena, Montana (see illustration), the Yellow Aster, at Randsburg, California, and the Treadwell, Alaska. Those mentioned have mined a very large amount of ore by this method, and afford excellent examples of this practice. We consider the general method in vogue at the Yellow Aster superior to most others, as more attention has been given to the details of preparation and the result is eminently satisfactory.

Cheap mining is done at all of the places mentioned by the "glory-hole" method. The term glory hole originated at the Treadwell mine, on Douglas Island, Alaska. So many miners employed in the great open cuts were killed by tumbling down the cliffs, or were knocked over by falling rocks, that the cuts were called glory holes, as, presumably, the poor fellows went to glory numerously and suddenly. Ordinarily, the open-cut method of mining is not excessively dangerous. Often the greatest element of danger is the bulldozing of large rocks, and the danger from this source is increased by the insane practice of placing rocks of 5 to 10 pounds weight on the charge of powder and fuses "to keep them from being jarred off." At times of bulldozing, the air is filled with rock fragments for several hundred yards, and men are frequently hurt who considered themselves at a safe distance; in most cases they would be safe from blasting, but it is the small loose rocks that constitute the greatest menace.

The sketch, Fig. 67, represents partly the present practice, and in part the prospective method of mining at the Yellow Aster mine, at Randsburg, California. The upper half of the drawing is intended to illustrate the method at present in vogue there, and the lower part the method that in all probability will be introduced later, when the large open cut is extended to greater depth, about 115 ft. lower than the floor of the cut where the principal out-door operations are now being carried on.

At present the ore broken down in the faces of the great cut either falls directly upon or near grizzlies covering mill holes, conveniently disposed at various points about the great excavation, or it is shoveled at the base of the ore face into cars and trammed to the nearest mill hole, through which it descends by gravity to the loading chutes situated on the level below. In order to continue the method of mining by open cut, it will soon become desirable, if not necessary, to carry excavation below the present level, called the Trilby. On the Rand level, which is the lowest adit level of the mine at this time, trains of cars, drawn

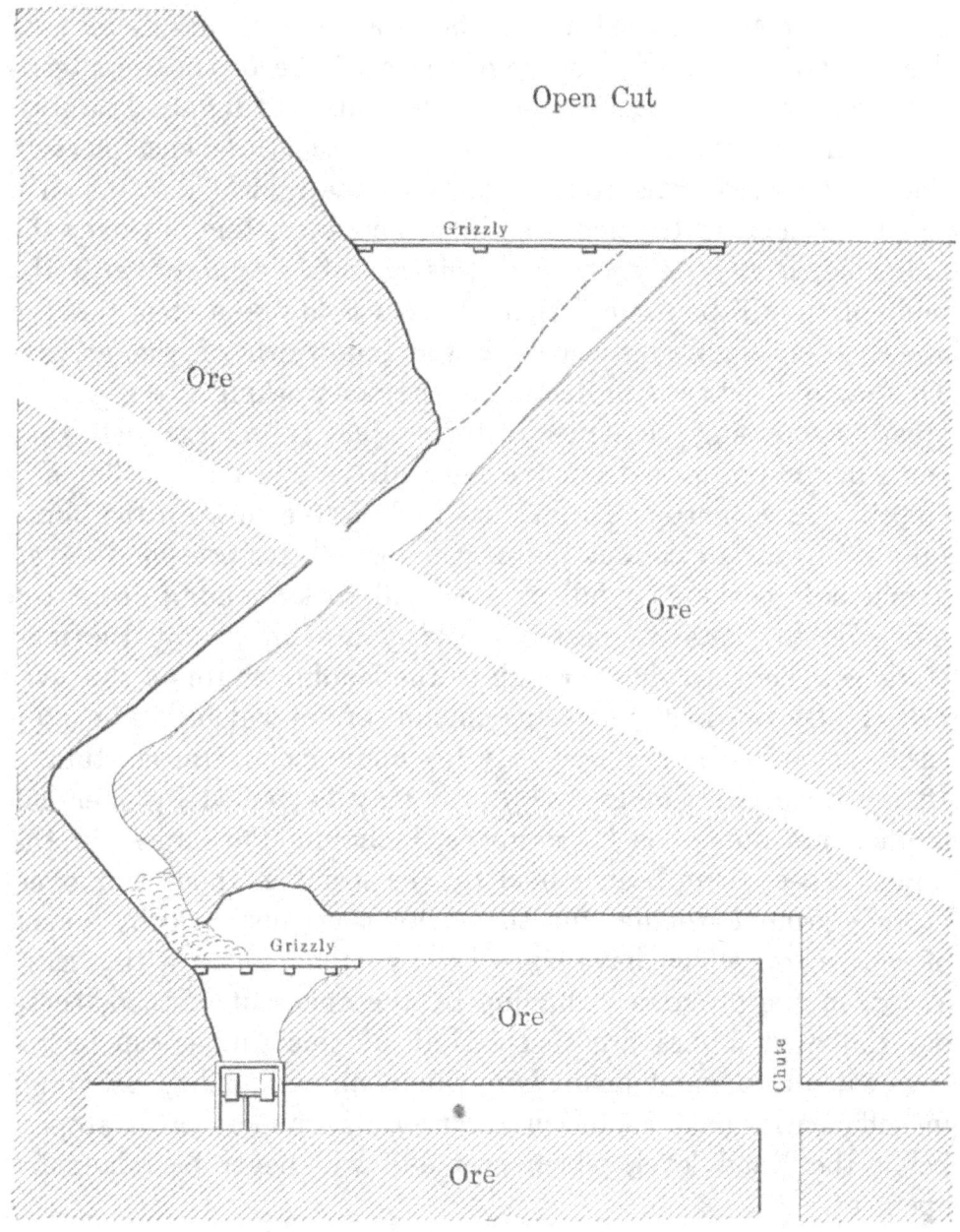

Fig. 67

by a motor, enter the mine and by a rudely circular route pass a series of loading chutes which connect by mill holes with the open cut on the Trilby level and with certain underground stopes on the Trilby level, and at others intermediate between that and the

Rand level. To facilitate the scheme of continuing the open workings below the Trilby level, it is proposed to cut raises about 30 ft. high at various points and to connect these by short levels with the existing mill holes, as shown in Fig. 67, or with raises to be cut later and utilized as mill holes at such points as may be deemed advisable. The enormous size of the ore-bearing territory has not only made this scheme feasible, but highly desirable.

It will be observed that at the end of the short level connecting the manway from the Rand level with the mill hole is shown a small chamber, in the floor of which, and extending out over the mill hole, is a grizzly. Where this chamber connects with the mill hole it will be noticed that the roof is low — a mere passage about 4 ft. high, but which is the full width of the grizzly. This low arch acts as a protection to the men working at the grizzly, greatly decreasing the danger of their injury from large rocks that come bounding down from the level above when the millhole is empty. As a matter of course, it is desirable to keep the chute full, but owing to various causes it is not always possible to keep a mill hole constantly full of ore. All rocks coming down the mill hole too large to pass the grizzly are broken by the men stationed there, passing through to the loading chute on the level below. By this method a large amount of ore will be made more easily accessible at reduced cost for mining and transportation. In the huge block of ground which may be mined by open-cut method are numerous large stopes, including the largest in the mine. These have been mined by the usual methods employed in underground stoping, and the excavations sustained by means of square sets which have since been partially filled with waste. When, in the progress of mining by the open-cut method, these old timbered stopes are reached, all of this timber can be recovered, and a great deal of it will in all probability be found in sufficiently good condition for re-use in underground stopes, below the Rand level, which probably will never be mined by open cut.

The accompanying engraving (Fig. 68) of the great open cut in the Mount Morgan mine, in Queensland, Australia, illustrates this. Formerly the ore was extracted by underground methods, the stopes, as at the Yellow Aster, being supported by square sets. These are now seen exposed in the sides of the later open excavation, and is a picture similar to that which will eventually

MINING LARGE ORE BODIES 139

Fig. 68. — Open cut of Mount Morgan Mine, Queensland, Australia

be presented by the Yellow Aster, or any other mine, where open-cut methods follow underground stoping and timbering by square sets.

At the Homestake mine, which extends for 10,000 feet along a gold-bearing mineral zone of prodigious size in the Black Hills of South Dakota, the earliest mining was by means of open cuts. These cuts, 8 or 10 in number in the early days, have in some cases been extended until they met, forming very large excavations, at Lead, Terraville and Central City. The largest of these, however, is the main Homestake cut at Lead. This cut, in the spring of 1878, was a hole about 60 ft. long, 40 ft. wide and not over 40 ft. deep at the deepest, up-hill side. This cut is now over 1500 ft. long, 300 to 500 ft. wide at the top, and in places more than 250 ft. deep, below the upper rim. An enormous tonnage of ore has been extracted from this and the other great open excavations of the Homestake system, at a remarkably low cost. On the east or hanging-wall side is an immense mass of barren porphyry. The long-continued excavation, both on the surface and in underground stopes, has caused acres of this overburden to crack and settle. The constant movement of the great rock mass has caused stresses to be exerted on the rock, and as a result it is shattered into millions of angular blocks of varying size. A portion of the porhpyry exists as a thick sheet, being part of a laccolith which formerly overlaid the outcroppping ore body. The remainder of it occurs as intrusive dikes coming up from below. As a result there is an intermixture of ore and barren porphyry near the surface, both rocks occurring in large masses. Owing to these facts, the porphyry has been utilized as filling for the immense underground stopes. It is broken as above described, by the shifting weight of the rock mass itself, and is constantly on the move downward toward the mill holes which connect the great cut with the underground workings.

The size of the ore body from the 500 ft. level to the 1100 level is unusual, averaging over 500 ft. in width. Its width below the 1100 level is as yet unknown. The excavation of an ore body of such tremendous size at minimum cost necessarily requires that the several important factors having a direct bearing on expense be most favorable. These conditions are found at the Homestake, and the management of that great property has introduced mining methods, both at surface and underground, which may be followed to great advantage in many other places where conditions are in any way similar. The fact of the intermixture of ore and waste in large blocks in the upper portion of the mine makes it possible, and even necessary at times, to send all ore, or all waste, to the extent of a thousand tons, more or less, through any particular mill hole. For several days men in the cuts will work down waste, which, passing from level to level, finally finds lodgement in some underground stope. After several days, possibly, a large amount of ore may have become available at this mill hole, and then only ore is sent down, to be drawn off at a convenient level, trammed to one of the several shafts, hoisted, and sent to mill. The condition has been such for several years past, in the great open cuts of the Homestake, that it is nothing uncommon for two men to work down 300 to 400 tons of ore daily.

The system upon which all open-cut mining work should be laid out and the work prosecuted should contemplate from the beginning the mining of a large tonnage of ore daily with the least possible amount of handling. Topographical situation and the character of the rock are important factors. A rock that readily breaks up into comparatively small pieces may be more cheaply mined than that which is hard and tough, coming down in large boulders, and requiring subsequent block-holing or bulldozing.

The Churn-Drill or Jumper

Methods of mining in open cuts differ according to the conditions under which the work is done. In many cases all the material broken is valuable ore. In others the ore is mixed with waste. Occasionally these — the ore and the waste — are so intimately mixed that sorting is commercially prohibited. In other instances the ore and the waste, though intermingled, both occur in such large masses that each may be mined separately, as in the Homestake cuts, as previously explained. Again, ore and waste are not intimately mixed, but occur in such manner that by blasting carefully the two may be kept separate to a great extent. Where this latter condition prevails, the holes are drilled in the usual manner by hand, or with machines. The drill holes may then be from 4 to 6 ft. deep, charged and fired in the usual way. Where, however, there is no need of such careful procedure, and there is no objection to breaking down hundreds, or even thousands, of tons at a single round of holes, the churn-drill or jumper, as it is often called by English miners, can be employed to great advantage. This method of drilling holes for blasting may be employed with equal usefulness in quarries where the rock is being broken for macadam, concrete, or other similar purposes, and where the shape or size of the rock broken is of no consequence, the chief thing being the amount that may be thrown down with a single shot.

The following is a description of the method employed where the rock was tough, but not particularly hard. The holes were started by hand-drills of the usual kind, the bits being $2\frac{1}{8}$ in. wide. These holes were drilled 5 ft. deep with double-hand hammer, when the drillers moved on to another place and started a second hole. One man with a 12-ft. jumper (churn-drill), made of $1\frac{1}{2}$ in. gas-pipe with drill-bits welded to each end, began to deepen the 5-ft. hole. When deep enough, a 16-ft. drill was substituted for the 12-ft. drill and two men then handled the drill. The third drill was 20 ft. long, and when this came into use three men were put on the drill. Following the 20-ft. drill came the last one, from 24 to 26 ft. in length. With the longest drill three men still constituted the drilling crew. The widest bit on the churn-drill was $1\frac{7}{8}$ in., the narrowest $1\frac{3}{8}$ in.

The bits should be forged strong — it is a mistake to make them too thin, for the corners are likely to be broken off. The temper must be low — dark straw color to blue — or the corners may check and break, and it is difficult to remove broken pieces of bits from a deep hole. The best way to get them out is with a strong bar magnet, attached to a cord, which usually secures the desired fragment in a moment. Miners often waste much time in "spooning" after a bit of broken drill.

Clean the holes down to 4 ft. with the scraper. Deeper holes should be cleaned with eye-bars of $\frac{3}{8}$-in. iron. A strip of cloth or rope-yarn should be run through the eye. This, if churned up and down in the hole, becomes loaded with the muck, which is then drawn out of the hole and the superfluous part stripped off by running the rag between thumb and fingers. The process is repeated until most of the muck has been removed, when drilling may be resumed. The swab of rope-yarn must not be too long, or it may double up and cause trouble by jamming in the hole.

When the hole is finished to the desired depth, the chamber is started by shooting two sticks of 40 per cent., followed by 6 or 8 sticks of No. 1 (70 per cent.) dynamite; third time by 12 to 15 sticks No. 1, and the fourth time by 20 or more sticks of the same. The succession of these operations, and amount of powder employed and number of times, must be determined by the character of the ground and the way it breaks, as shown by actual experience. In some fairly easy ground two or three rounds of 20 sticks of No. 1, after the preliminary shots with Nos. 1 and 2, are sufficient, but in hard and tough ground it may be well to shoot 3 to 5 times with 20 or more sticks of No. 1, before the chamber is sufficiently large to receive the necessary amount of "low" dynamite or black powder.

When beginning the operation of "springing the hole," put down two sticks of No. 2 and follow it with the tamping stick to be sure it is at the bottom of the hole. Insert the detonator in a full stick of powder. Use a 30-in. fuse. Spit the fuse, allow it to burn about two inches, and then close the end by pressing the burned portion with the fingers. This has a tendency to exclude any water that may seep into the fuse and extinguish the fire. Drop the fuse and primer into the hole and get out of the way of small rocks that may be blown up out of the hole. After springing the hole, pour water into it to extinguish any fire in any piece of fuse tape, before pouring in black powder or low dynamite. In a 20-ft. hole that has been properly chambered it is customary to use 200 lb. of Judson 5 per cent. "low" powder. It comes loose in 50-lb. cases. This quantity may or may not be sufficient. Experiment only can determine this.

The distance that a series of holes may be drilled back from the collar varies with the rock. In some cases 10 to 12 ft. is as heavy a burden as may safely be placed on the hole; in other cases it may be as much as 18 to 20 ft. back from the face. A single hole drilled 20 ft. deep and 20 ft. back from a vertical face is theoretically calculated to break about 600 tons of ordinary ore, but a series of four holes placed 30 ft. apart, 20 ft. back from the face and drilled 20 ft. deep, is likely, in fair ground, to break from 4000 to 6000 tons of rock — a very much higher average than can be accomplished with a single hole 20 ft. deep with a 20-ft. burden. Moreover, a series of such holes is more likely to accomplish desired results than a single hole. Churn-drilling is recommended as a cheap method of breaking rock where the conditions are suited to this class of mining practice. It is an inexpensive method where the conditions are right and when the work is done by experienced men who are not afraid to work. At mines where there is a proper air-compressing plant and good drills, the holes may be drilled by machine to a depth of 10 ft. and even more, to advantage, before employing the churn-drill, and with suitable machines the holes may be drilled to the depth of 20 ft. or more. Such machines, however, are larger than those ordinarily employed in metal mining. The question of economy of machines over hand work must be determined by the management. In churn-drilling the personal equation is a very important factor.

The Steam Shovel in Open Cuts

In no part of the world is mining carried on on a larger scale or at less expense per ton than in the iron mining regions of Minnesota and Michigan. In these districts many millions of tons of ore are mined annually at a surprisingly low cost. Both surface (open cut) and underground methods of mining are in vogue, but by far the larger part of the ore is mined in the open, in which the steam-shovel is a most important factor. The procedure in the Mesabi Range is simple. The glacial drift, or "overburden," is stripped from the ore with steam-shovels. The thickness of the drift removed may be only a few feet, or as much as 85 ft. The average is between 20 and 40 ft. Among mining men the expression is common that it pays to strip as great a thickness of drift as there is ore beneath. However, factors other than thickness of the ore beneath frequently determine the amount of drift that it is advisable to attempt to remove. The total amount of drift that has had to be removed from some of the large deposits is very great, numbered by millions of cubic yards. When the surface of the ore, or part of it, is stripped, standard-gage railroad tracks are built out on the ore deposit and steam-shovels make a cut through the ore. The ore of the Mesabi range is soft and may be mined in this manner by steam-shovel. In the first cut the ore is either thrown to one side, or is loaded directly on to cars on a parallel track. After the first cut the shovel is set over against the bank and another slice taken off, and loaded on to cars run in the cut already made.

When, by a series of cuts or slices, the bank or bench is carried back far enough, work is begun as before on a lower level, and in time this is followed by cuts on a third and fourth level, carving the deposit into a series of banks or terraces, at several levels, against any or all of which steam-shovels may work, giving access to a great variety of ores and making possible a large output in a short time. In some of the mines the cut is started near the middle and the work is carried forward toward each side. In others the first cut is made in spiral form, leaving a bank in the middle, so that subsequent cutting goes on both toward the center and toward the periphery of the deposit at the same time. While the ore is soft, it is usually too compact to handle economically without blasting, so a small amount of this is done for the purpose of shaking up the ore. The system of trackage varies greatly in the several mines, depending upon the distribution and character of the ores and the shape of the deposit. The mill hole is also made use of in some of these mines, but it is a method not generally adopted in the Mesabi region.

The glacial drift is removed as in an open-cut steam-shovel mine. A shaft is sunk in the adjacent wall rock to the level of the bottom of the deposit, and drifts or cross-cuts are run out through the ore. Raises or chutes without timber are sent up from the cross-cuts or drifts. By blasting, the ore is then loosened at the surface and pushed into the mill holes by men. It is drawn into cars stationed at the chutes, trammed to the shaft, and hoisted. Another practice is to dump the ore directly into the mill holes by means of steam-shovels instead of by manual labor. This method has been found satisfactory.

In a comparison of methods — steam-shovels operating along benches or terraces, the mill-hole system, and underground mining, the advantage in cost per ton lies with the steam-shovel. By that method all of the ore may be recovered, the work may be performed largely or wholly in daylight, no timber is required, and the number of men necessary for the handling of a large tonnage within stated time is reduced to a minimum.

So successful has the steam-shovel in open cuts become that the method has been extensively introduced in the metal mines of the West and abroad.

The first to employ the steam-shovel in a metal mine in the West was the Granby Copper Company, in British Columbia.

146 TIMBERING AND MINING

FIG. 69. — Mining with Steam-shovel at Ely, Nevada.

This was several years ago. Since then the steam-shovel has been introduced with success at Bingham, Utah; at the Cactus mine, in Beaver County, Utah; at Mount Lyell, Tasmania; at the great copper mines of the Ely district, Nevada (see Fig. 69), and elsewhere, and there are scores of other places where they may be employed to excellent advantage, particularly in the removal of overburden from ore deposits which may then be mined by steam-shovels, or by mill-hole system.

Chapter XIV

THE OVERHAND AND UNDERHAND METHODS OF STOPING VEINS

The removal of ore from veins or deposits in underground workings is commonly known as stoping, and the excavations made in this manner are called stopes (originally steps), the ore usually being removed in horizontal slices, leaving a series of "steps" in the vein. Stopes vary greatly in size and shape, depending partly on the position and form of the ore deposit and partly on the character of the ore or the surrounding country rock. A vein of hard ore in firm walls can be stoped in a very different manner from a soft and crumbling vein, or one where the walls are soft or otherwise insecure. The size of the vein or deposit and the angle or pitch of its walls will also often have a very important bearing on the method of stoping adopted.

Stoping is of two kinds, distinguished as underhand and overhand. Underhand stoping, which is the method of removing ore by downward steps, is almost exclusively confined to veins of small width and was at one time very generally practised, but is now only rarely seen, and is then usually the result of an endeavor to remove ore quickly from a vein without the necessity of first doing the development work otherwise necessary to get under the ore body. It is an expensive method and is attended by many disadvantages. All the water coming out of the vein or walls runs down the several terraces of the stope, keeping it wet and mucky. All ore has to be handled with greater care and often picked by hand. Underhand stoping may be justified in small veins of rich ore, and where the geological conditions indicate that the ore shoot does not go to great depth, but generally speaking underhand stoping is inadvisable. Several years ago the writer visited a mine and, accompanying the superintendent down the shaft, found the first level opened 125 ft. from the surface. Beneath the level an underhand stope had been started fully 20 ft. wide and about 40 ft. long. It was being timbered in square-sets. Having never seen anything like this previously, the superintendent was asked why this method of mining had been attempted. His reply was that he had a smelter of 100 tons capacity to supply. This was a desperate effort to meet the requirements of a metallurgical plant which had indiscreetly been built before the mine had been sufficiently developed to warrant a smelter. Within 30 days from the date of the visit above mentioned both smelter and mine were closed.

Underhand stoping was formerly very popular with Cornish miners, for various reasons, but is less in vogue now, as the more direct and economical method of overhand stoping is almost exclusively practised. Underhand stoping may, however, be proper in a few instances where it is desirable to remove ore from the surface downward for a few feet, say 20 to 40 ft., and where the hoisting is to be done by means of a windlass or whim. A shaft is sunk to the depth to which it is intended to carry stoping, or a little deeper, in order that a sump may be formed at the bottom to collect the drainage, if there be any water present. The stoping then begins by mining a shallow open cut about 8 or 9 ft. deep, along the surface, carrying the work in each direction from the shaft along the vein. The ore removed is taken in a wheelbarrow to the shaft and dumped on an apron, which delivers it to the bucket, or the bucket is hauled in onto the level and the ore shoveled directly into it from the accumulated pile. When the work in this cut or trench has advanced 20 ft., or thereabouts, a second cut may be started 8 or 9 ft. lower, and carried forward in the same manner. If the first cut has not been carried to the limit of the ore shoot, stulls must be placed at the level of the floor of stope No. 1, and these covered with lagging, forming a strong platform over which the miners working in the upper cut may carry their ore, and on which may be accumulated any waste or other debris resulting from mining in stope No. 1. When stope No. 2 has been advanced sufficiently, stope No. 3 may likewise be started, and later No. 4, each being driven from the shaft toward the limits of the ore shoot. It is only in rare cases that this method of mining is seen in the Western United States, as it involves too much work for the results accomplished, and it is only in exceptional cases that its adoption may be recommended.

In Cornish mines, where underhand stoping is practised, the usual method is to sink the shaft, drive several levels, and connect these several levels with a series of raises, some of which are utilized as ore passes. The ore is stoped by the underhand method, conveyed from the several working faces to the nearest mill hole, into which it is dumped, to be drawn off into a car through a chute built at the foot of the raise. The mill hole is sometimes cribbed and sometimes planked, open places being left at each floor so that the ore may be dumped into the mill hole. The ore must be drawn off about as fast as it is thrown into the mill hole, as it cannot be allowed to accumulate to a higher level in the mill hole than the lowest working floor in that stope. Overhand stoping is the most commonly practised throughout the metal mining regions of the world. A treatise on stoping is in itself a large undertaking, and we have no intention of describing its many variations, referring only to those methods most commonly in use and which have been found to be the most satisfactory.

The methods of stoping may be divided into several kinds: Stoping with the use of simple stulls to support the walls; stoping without timbers, by allowing two-thirds of the ore to remain in the stope until the ore is all removed, when it may all be drawn off; stoping by means of the square-set system of timbers; by room-and-pillar method; by the caving system; or by some other one of several methods not commonly practised because they have been evolved to meet peculiar or unusual conditions not usually found in mines. The simplest method of stoping by the overhand system is in the removal of a vein of clean ore, where the vein is of moderate width — say less than 16 ft. This is accomplished in several ways. In former years it was the practice to drive a level along the vein, placing heavy drift sets, if the walls and character of the vein demanded it, and laying thick lagging on the caps. The ore overhead was then broken down onto the lagging and shoveled into chutes built at intervals — usually of 30 ft. — along the level. As stoping proceeded upward, stulls were placed and these were covered with platforms on which the miners stood when at work. The ore was sent down to the chutes on the level through ore passes (sections in the stope cribbed with timbers) and drawn off into cars. This method eventually leaves a large, open excavation in the mine, the ore all having been drawn off, and only the stulls remaining to support the walls, the platforms of lagging having been removed.

Fig. 70 illustrates the method of placing stulls in a stope, the engraving representing one that has been carried up to the surface from below.

In many veins of high inclination the conditions are such that

Fig. 70. — An Open Stope in a California Mine

stoping may be most conveniently and economically carried on by breaking all the ground in the stope, drawing off only about one-third, or only so much as will permit the miners head room while standing on the broken ore, to continue the work of mining overhead. When this is done, the ore is drawn off through chutes built at intervals of 30 to 40 ft. along the level. On the level, drift sets may be put in, covered with heavy lagging, or, if the walls be hard and firm, stulls only need be employed. The loading chutes are built in the usual manner, either attached to upright posts or built in between two stulls. As the ore is broken down, the miners break up all large pieces, either by bull-dozing or block-holing, and finally, if necessary, with hammers, until no piece remains that will not pass through the loading chute. All the rock broken in the stope may be drawn off through the chutes without shoveling, except that which remains on the lagging of the gangway sets when the "angle of convenience" has been reached near the bottom. When the rock ceases to run by gravitation to the chutes, men go into the stope and shovel the ore lying on the lagging into the chutes. If this be too dangerous or for any reason impracticable, an intermediate chute may be put in midway between two regular chutes and most of the ore recovered. At each end of every such stope must be arranged manways for the entry and exit of men and for the purpose of maintaining a free circulation of air. But one of these openings need be provided with ladders, though in many mines it is the custom to hang a rope from a stull in the other in anticipation of an accident to the main manway. The above assumes that the vein consists of clean ore, or that the waste is present in such small amount as to make it a matter of relatively small consequence.

Very often waste occurs with the ore to such an extent as to form a very considerable part of the vein material, and often, too, the walls are not as firm as could be desired, large slabs being loosened and falling upon the broken ore. These must either be supported with stulls or taken down to insure the safety of the miners. When such conditions obtain the ore must be sorted from the waste and sent down to the loading chutes on the level below, through cribbed mill holes that are carried up in the stope as the work progresses upward. It is the best practice to keep these mill holes full of ore, even if the amount broken is comparatively small. By doing so the danger of men falling into them is obviated, and no large pieces of waste can drop into a position where it is disadvantageous to break it up. Often, while there is considerable waste encountered in the vein, or the waste from the walls is so troublesome that the method of mining just described is adopted, still the amount of waste is not sufficient to keep the stope filled up close enough to the back to keep the miners within reach of the ore. In such cases it may be expedient to break enough of the walls to supply the necessary filling. As a matter of course, the breaking of all the ore and allowing it to remain in the stope has its advantages over this latter method, which requires sorting, cribbed mill holes and other expenditures of labor and money that in the former case are unnecessary, but we have not always either clean ore or firm walls. Which of the two systems is adopted must be determined by the judgment of the superintendent. In some instances where waste occurs, but is not particularly abundant, the first method is employed, the greater part of the waste — the larger pieces — being thrown out by hand at the loading chute, at the surface chute, at the rock-breaker, or wherever it may show itself. Not infrequently conditions in different parts of the same mine are so unlike that the methods employed in the various stopes differ, according to the existing conditions.

Chapter XV

UNUSUAL METHODS OF STOPING ENFORCED BY SCARCITY OF TIMBER

In desert regions the scarcity of timber is an important feature in mining economics and has led to the introduction of unusual methods of ore recovery. These methods may also be employed in other mines where timber is not scarce, and may in some cases be found to possess advantages over old methods.

One of the most practical and satisfactory methods of stoping in veins of normal width — 5 to 20 ft. — is to drive a gangway in the vein, put in raises at every 30 ft. and at a height of 10 to 20 ft. above the back of the main gangway (depending on the size, character and condition of the veins, and its walls), and to open a level by driving an upper drift, from which stoping proceeds upward. This leaves a solid block of ore between the back of the main gangway and the floor of the stope. Of course, this method anticipates that the vein is firm and that no timber is required to support the back of the main gangway. This block of ore is only penetrated by the raises cut for ore passes and manways. The ore is stoped in the usual manner and only enough drawn off to keep the miners within reach of the back of the stope. The mill holes are those cut through the block of ore, each being provided with a chute in the gangway below. The end raises are maintained as manways and for ventilation, and may require some timbering.

The Black Mountain mine, at Cerro Prieto, Sonora, Mexico, is being worked in this manner, very similar to the method employed in some parts of the Treadwell mine, Douglas Island, Alaska. The mines of Zaruma, Ecuador, also afford an interesting example of special methods, described by Mr. J. R. Finlay in the "Transactions of the American Institute of Mining Engineers" (Vol. XXX, page 248).

The method followed there was suggested by the practice at an iron mine at Tower, Minnesota. The excellence of the scheme led to the adoption of a modification of the idea at the Eagle-Shawmut mine, near Chinese, California, by the superintendent, Mr. Charles Uren. Wages at Zaruma are 50 cents a day (gold) for common labor; 60 cents to $1 for native miners; $1 for Jamaica negroes; $2 for Italian miners. American mechanics and miners receive $140 per month and upward.

Mr Finlay thus describes the method of mining introduced by him at Zaruma: "The first ore body opened by the present company was on the 'Abundencia' vein. The ore was not very hard except in spots, and was accompanied by an exceedingly soft and treacherous foot-wall. It was evident that, whether timber were used in the stopes or not, they would have to be filled with rock. The only other alternative was to slice the body from the top downward. The latter method may yet have to be adopted near the surface, but it involves great difficulty in keeping open the raises, which are necessary, both for ventilation and for furnishing future rock-filling for the lower levels. The method of mining actually adopted was essentially the filling system used at the hard ore mines of the Minnesota Iron Co., at Tower, Minneapolis, with the exception that the fills, and of course the stopes, were made sloping instead of horizontal. In the first ore body attacked, three raises, each from 200 to 300 ft. high, were made along the foot-wall to the surface, coming out at the bottom of the deep open cut of the old Spaniards, from the sides of which any quantity of rock could be milled down into the mine at a nominal cost. These three raises were 130 ft. apart. The tunnel was run, not in the vein itself, but at a safe distance (about 20 ft.) in the foot-wall. Cross-cuts were driven to the vein every 65 ft., so that there was an entrance to the stopes at each of the raises, and also midway between them. Stoping began by cutting out the vein to its full width, and about eight feet high, on the main level. Tracks were laid in the cross-cuts between the raises. When the bottom was cut off, the vein was also beaten away as far as could safely be done at the bottom of the raises. Then a lot of waste rock was thrown down each raise, which nearly filled the openings thus made. This filling was allowed to lie nearly at the angle of repose, and upon the sloping sides of the pile slabs were laid, to keep the ore from mixing with the rock. Then another slice was taken off as high as safety permitted, and the operation was repeated. Cribbed manways were, of course,

156   TIMBERING AND MINING

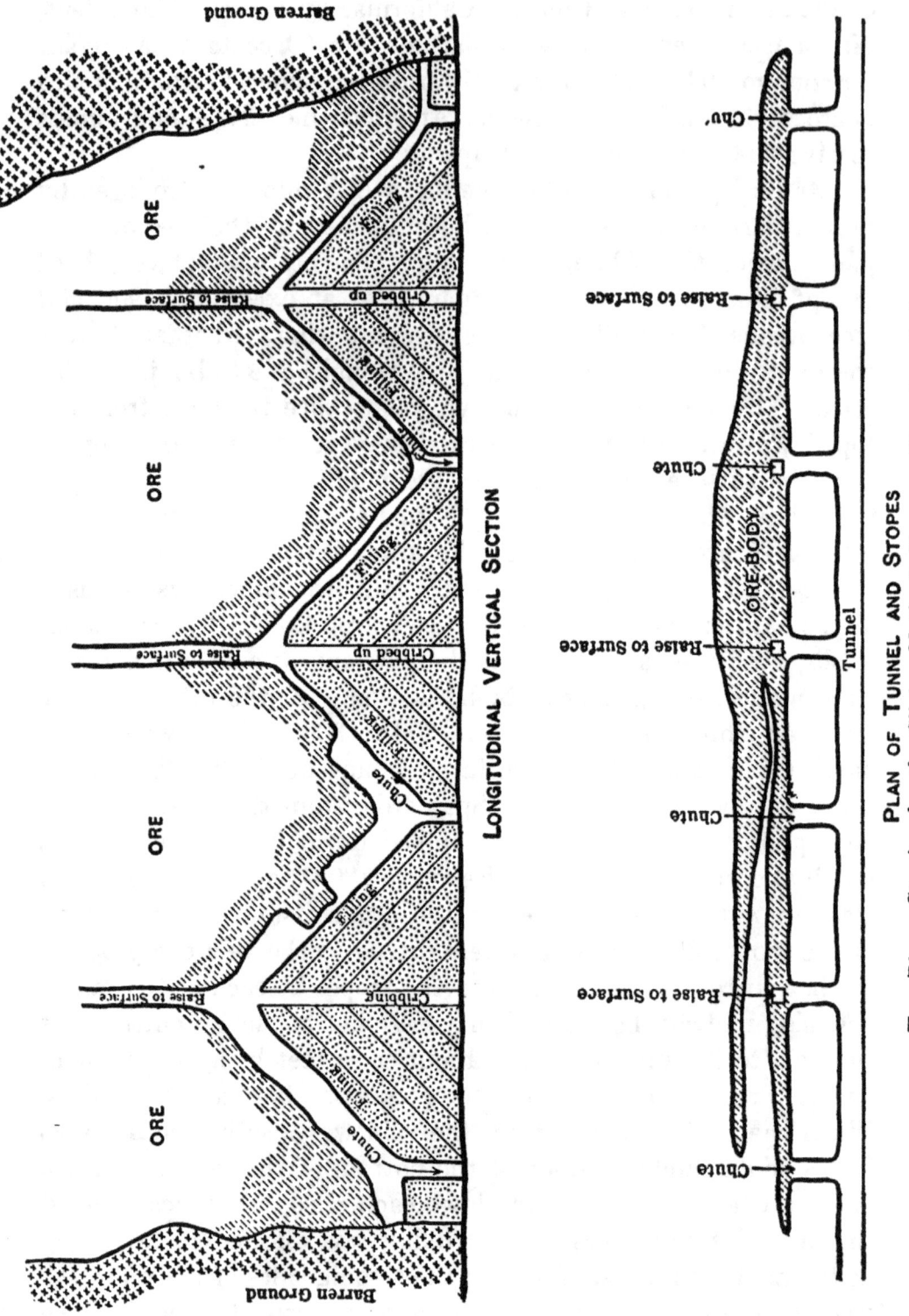

Fig. 71.—Stoping by the Slicing Method, Zaruma, Ecuador, S. A.

SCARCITY OF TIMBER

carried up through the filling, to preserve the raises for future use. The slopes on each side of the filling raises soon came together, half way between the raises, or at the intermediate crosscut. At these points chutes were put in, and carried upward by cribbing as the stoping and filling proceeded. Fig. 71 illustrates the mining method adopted at Zaruma.

"This method of mining has the following advantages: (1) It requires but little timber — an important consideration where timber is scarce. (2) It allows of filling and tramming without re-handling, and, therefore, at a minimum expense. (3) It allows the fill to be made very close to the back, because it is not necessary to leave standing room on the slope for men to stow the dirt. This is a very important consideration where the walls are soft.

"The system works well. The slope-filling feature of it was invented by Mr. Thomas Huddlestone, mining captain on the property. The ore costs, for labor, supplies and superintendence, about 70 cents a ton, delivered at mill. This figure relates only to the operation of the stopes, and does not include the large amount of development work done."

It is not quite clear what keeps the slabs in position on the sloping sides of the fill, nor how the miners find a secure footing by this method.

Chapter XVI

STOPING IN FLAT OR LOW-LYING VEINS

In veins which lie quite flat, the character of the ore and that of the overlying wall must necessarily determine the method of mining that may be adopted with safety and economy. In some localities flat deposits of ore lie at the surface, or have very little overburden that must be removed. These conditions afford an opportunity for very cheap mining. In the Black Hills of South Dakota are numerous deposits of this character. The ore occurs in and on the Cambrian quartzites, and in a number of cases erosion has removed the later deposits and the bed of ore lies uncovered at the surface. Among the several mines of this type, that at the Wasp No. 2 mine, on Yellow Creek, is perhaps the most noted example. There open-cutting in these flat veins has been reduced to a science. The practice there is usually to drill a single hole from the surface to the bottom of the ore bed, about 20 ft. This hole is set back 30 to 40 ft. from the face. It is chambered repeatedly, as described under the head of churn-drilling. When, in the judgment of the foreman, the chamber is sufficiently large, several hundred pounds of black powder are charged and the blast fired. The result is that from 5000 to 6000 tons of ore are lifted slightly, settling back into position again. None but a close observer would notice any material change in the appearance of the ore after the "big shot" had been fired, but the entire mass is thoroughly shattered, and for a month the miners have only to shovel away the broken rock, the larger pieces being bulldozed or block-holed.

Where the flat veins extend beneath a mountain, all opportunity for adopting the method just described is lost, and the vein must be attacked in a different manner. The system of mining will in such cases be very largely influenced by the character of the ore and of the walls. Where the vein is thin (a foot or two), usually very little timber is required, the waste rock filling the entire space behind the miner. Where the mineral deposit is thicker and timber is necessary, various methods are pursued. Some ground stands well by simply leaving pillars of mineral. In other cases a series of upright posts and breasting caps will sustain the roof, the posts being placed in rows directly back of the workmen and as close to the face as necessity demands. The foot of the post rests either directly on the rock floor or upon a block of wood or piece of heavy plank. The posts are forced into position by driving them up with heavy hammers. Care must be taken that these posts are so placed as to receive the weight of the roof directly, and not at an angle. These timbers are set in lines standing in two, three, or four rows back from the face, the waste being piled behind as the work advances. In this manner, by exercising care, many sticks of timber can be recovered before the weight settles so heavily on the refuse rock as to render it impossible to remove them. Some flat veins make little or no waste. It is then necessary to follow the "pillar and stall" system of extraction, considerable blocks being left to sustain the roof. Posts and caps are used in this system also. Frequently the caps reach in a continuous line from post to post, joining the next set, the ends of two caps resting on a single post, the combined sets being a hundred feet or more in width. Large timbers thus placed will support great weight, but if small rocks fall from the roof, lagging also must be employed. This is the system much in use in California drift-gravel mines.

When a vein lying nearly or quite horizontal, and making no waste, is to be mined, a drift should be run along the lowest portion of the deposit, this point having been reached by incline or shaft. The work advances towards the surface, good-sized pillars being left to sustain the roof. If timber be necessary, it is put in place in the manner required. The work having advanced sufficiently toward the surface, the pillars may now be cut out at the back end, while the work progresses as before. As the pillars are removed more timber must be put in, or waste from the surface must be piled in cribs of timber built in the workings already made. Usually some timber can be recovered in this way, and the caving of the roof, after the complete removal of the ore or mineral, does no harm. The main gangway should be substantially timbered, if necessary, as it is desirable to keep it open to the lowest working level at all times.

The "long wall" system of extracting ore is usually carried from the surface inward, a main gangway having been first driven ahead to a connection with a ventilating shaft, when possible. All the ore is removed at once, the waste being thrown back of the miners, who carry the breast forward with the center considerably in advance of the sides, the excavation being in form somewhat like the letter A, with the apex forward. The waste is thrown into the center to support the roof, while the side passages permit of a free circulation of air all along the face.

Veins of 10 ft. or more in thickness, which lie at an angle lower than 35° may be worked economically by sinking an inclined shaft in the foot-wall, 25 ft. beneath the vein, driving levels to and across the vein, cutting ore pockets at each level in the country rock between the shaft and the vein, and stoping the ore body in blocks. Filling may be obtained from chambers cut in the hanging-wall country, or it may be broken in an open cut on the surface and sent down into the mine in cars, an inclined shaft having been sunk in the hanging-wall for this purpose, from which drifts extend over the stopes. In lieu of this latter arrangement, the waste rock may be sent down in cars in the working shaft sunk in the foot-wall, and these run off at the level above that to be filled, and the filling run down into the stope below through mill holes provided for the purpose.

Chapter XVII

RAISES FOR CONNECTION OF LEVELS

THE driving of raises in all mines is important, and in many of them necessarily precedes the commencement of stoping. Ordinarily, raises are cut either by machine or handwork without any more timber than absolutely necessary to enable the miners to climb up to the top and renew drilling after a blast. Stout stulls are placed in hitches as the work proceeds upward, and the men stand while drilling on platforms laid on the uppermost stulls. When ready to blast, the platform is piled up to one side, and the rock broken by the shots falls down the raise, to be drawn out at the bottom, where a chute is usually built, though in many cases the rock falls to the floor of the level below and is shoveled into cars and trammed away. There are instances, however, where it is desirable to divide the raise into two compartments, for the purpose of making it a permanent ore pass and also to afford a means of ventilation. In this case a line of stulls is set up at or near the middle of the raise, at right angles to the angle of the raise itself, and on the ore-pass side is spiked 2-in. planks. In the opposite compartment is placed a ladder to facilitate the passage of men from level to level. It will be understood that a raise of the character here described is entirely different from the mill holes carried up in a stope.

Sometimes it is desirable to timber a raise — in fact, to so construct it that it may, when finished, be used as a working shaft, or it may be desirable to connect it with a shaft sunk from the surface, or from some level above. Rapid shaft building, as it may be called, is often accomplished by driving a number of raises simultaneously from several levels, the series being finally connected, making one continuous shaft. It is scarcely necessary to say that the greatest care on the part of the mine surveyor is necessary when such work is undertaken, as hoisting shafts require absolutely straight alinement, and no off-sets are permissible. A competent mine surveyor can lay out the work with such accuracy that the several sections of the series of raises will meet within a small fraction of an inch. It is, however, wise to make raises of such liberal size that any small discrepancy in alinement may be corrected by changing the thickness of the guides a little. Contemplating a series of vertical raises driven to connect with an existing shaft, each section will be a counterpart of the others. At the places selected for beginning the raise, cut out a station large enough to accommodate the loading chute and car tracks, either single or double, as required: for a single track of ordinary gage — say 18 in. — cut this station 12 ft. high and as wide and long as the shaft requires. Give sufficient working room on three sides of the car, at least, and space for the chute timbers on the fourth side. If the drift has been cut of liberal size, it is quite likely to afford room for all this without cutting out additional ground at the time. This station should be at one side of the shaft, not immediately below it.

When the raise has been started at one side of the drift, hitches should be cut at each end of the raise (or shaft, as we may call it, in anticipation of its purpose) to hold long, heavy timbers which are to serve as bearers, the same as in sinking a shaft. The size of the timbers may be of the same dimensions as those in the shaft sunk above, or may vary, in the discretion of the superintendent, with the character of the ground, it being permissible to employ smaller timbers in ground that is hard and firm than in ground that is soft and likely to cave. The bearers are to support the shaft timbers and also form a substantial back for the top of the rock bin and loading chute. Up to the time the raise has advanced 10 ft. all of the rock broken should be shoveled into cars from the floors and trammed away. Then the bearers should be put in place and firmly wedged. They must not only be level, but should conform exactly to the inside dimensions of the shaft or raise.

When the raise is up 10 ft. above the back of the station and the bearers are in place, the first set of shaft timbers should be placed in position. These consist of two wall plates and two end plates. The style of framing is left to the miner. Either the overlap or the dovetail corner is good. The latter is particularly applicable to inclined shafts, but may be used also in vertical shafts. When the lowest set of plates has been fixed in position and securely wedged to exact place, set up the posts on each corner, being careful that they rest squarely in the daps provided for them on the plates. On these four posts place the second set of wall plates; block and wedge them firmly, as before; put lagging in position all round, if necessary, and block it with rocks or wedges. The first shaft set is then complete. Build the (temporary) rock bin beneath one compartment Place a ladder on the floor of the station leading up to the first set.

This ladder may be placed in any desired position, clear of the track, to reach the first set. Two movable ladders must now be provided, to be used as the work progresses, one about 6 ft. 6 in. long, the other about 12 ft. Also make a number of permanent ladders about 16 ft. 6 in. long, to reach the height of three sets, when resting at an inclination of about 70°. In a vertical shaft, strong ladders, securely placed, are absolutely necessary. Never place ladders vertically when they can be set at an inclination. Put them in the manway compartment, on the side next to the dividers. This leaves the other end of the same compartment open for pipes, and for the passing of timbers, machines and supplies from place to place while the raise is being driven. At every third set, above the first, spike a 2 × 12-in. plank across the manway compartment, parallel with the dividers, and 22 in. from it. On the divider spike a piece of 2-in. lagging and on these spike two pieces of 2 × 12-in. plank side by side, parallel with the wall plates and at one side of the shaft. This leaves a manhole at every third set. The ladders should be placed with the foot resting on the platform 8 in. from the wall plate, and should extend at least 12 in. above the set at the upper end. It is also a good idea to provide a substantial, safe hand-hold a foot above the top of each ladder, as an aid to men who may be carrying tools, etc., up the raise. An iron dog 8 in. wide driven firmly into the lagging at the side of the shaft or a piece of 1 × 4 spiked to the posts, will be sufficient. The ladders should be placed directly above each other, inclining always in the same direction, so that when a man reaches the top of a ladder he steps to one side onto the plank platform and back to the foot of the next ladder, to continue his upward climb.

To hoist timber quickly and safely while the work of raising is in progress, a portion of the open space in the manway should be boxed in on four sides, providing a chute through which timber and other things may be hoisted with a block and tackle, or by other means. If this chute be made tight it will greatly aid ventilation — a most important matter, when raising, under any conditions.

Having described the method of placing ladders, platforms, etc., we may now return to mining operations. Assuming the raise now to have been cut to a height of 10 ft. above the bearers, and one complete set to be in position, place a short ladder from the first platform to the top of the set just put in. Over the manway compartment construct a heavy bulkhead of 8 × 8-in. timbers, leaving the rock compartment open. Spike lagging vertically on the four sides of the rock compartment. Two of the timbers of the bulkhead may be left out until ready to go below. A full round of holes of moderate depth should be drilled by hand or by machine. When ready to blast, if a machine has been used, lower it through the hole in the bulkhead and place it and other tools on a temporary platform on the set below. The miners spit the fuses on that side of the raise over the bulkhead, the holes having been pointed so as to throw the rock on that side as much as possible. After spitting the fuses, the men go down through the hole in the bulkhead, pulling the loose timbers in place and descending to the level below. When all the lighted holes have gone off and the smoke has had time to clear, the men return to the top of the raise and knock off two or three lagging of the rock compartment, so that they may climb into it and upon the bulkhead and its load of rock. The lagging is again spiked fast and the loose rock overhead barred down. All large pieces are broken to a size small enough to pass the chute door below and the rock is shoveled into the rock compartment, clearing the bulkhead. When the back has been cleared up, the remainder of the holes of the round are fired the men going below through an opening in the bulkhead as before, closing the hole after them. If the broken rock has filled up the rock compartment, a portion of it may be drawn off into cars and trammed away until the rock compartment has been relieved sufficiently to permit the miners to return to work through a hole made by knocking off the lagging, as before. Again the loose rock is barred down and the bulkhead cleared, and then the back is ready for another full round. When the rock bin has once been filled it is not necessary to exercise so much care in the pointing and loading of holes, as the danger of knocking out the chute timbers and the door is then practically at an end.

When sufficient headway has been made another set is placed in position and the bulkhead is moved up a set. The broken rock is never drawn off after blasting lower than the bottom of the last set placed, and the rock shoveled from the bulkhead and barred down from the back usually fills the rock compartment so nearly full as to afford ample protection to the timbers not protected by bulkhead. In this manner the work proceeds. After each round of holes is drilled the miners place their tools on a temporary platform built on the first set below the bulkhead, spit the fuses, and retire through a hole in the bulkhead, closing the hole after them. Returning, they break a way through the lagging into the rock compartment, drawing off as much rock as necessary to do so, and after clearing up proceed with the drilling of another round. When the work is well advanced, the long, permanent ladders may be put in place.

The question of ventilation is all-important, but may usually be solved by carrying up a pipe close to the bulkhead, or the timber chute built in the manway may be used for this purpose. About the only ground in which this method of driving a raise will not prove satisfactory is wet, clayey ground which would pack in the chute. In such cases it would probably be better to place a bulkhead over the rock compartment also, leaving open spaces of 6 to 10 inches between the timbers. This would act as a grizzly and retard to some extent the force of the flying rock when blasting. In such ground, however, it is generally better to raise with stulls, placing the permanent shaft timbers when the job has been completed.

Chapter XVIII

STOPING AND FILLING IN A MINE HAVING WEAK WALLS

In a former chapter was given a description of stoping in a vein of moderate width, first, in a clean vein, breaking all of the ore and drawing out only sufficient to allow the necessary head room for miners drilling in the backs. By this method the ore is drawn off into cars on the track in the level beneath. Second, mining in a vein in which considerable waste occurs, or where the walls are loose and shelly, coming down with the ore and unavoidably mixing with it. When this condition obtains, or the vein is so small that a portion of one or both walls must be broken to allow working room, mill holes are carried up in the stope, the ore is shoveled into these while the waste accumulates in the stope, keeping the men up to the back, or if there is too little waste for this, a means of providing it was suggested. In mining many conditions are found in veins, and the methods of meeting these conditions vary almost as much as the variety of conditions themselves.

Following is a description of mining a vein of iron ore at Soudan, Minnesota, written by D. H. Bacon, in the "Transactions of the American Institute of Mining Engineers," Vol. XXI, page 299. The method there described is somewhat different in some respects from those previously mentioned. It is thoroughly practical and readily applied where the conditions are in any manner similar.

"The iron ore deposits worked by the Minnesota Iron Company, at Soudan, Minnesota, occur in lenses 200 to 1000 ft. long and 5 to 80 ft. wide, and stand at an angle of 65 to 75 degrees, with a vertical height of 250 to 500 ft., other lenses occurring below. A number of the deposits were first worked as open pits, which in some cases were carried to a depth of 150 ft., when, owing to the weakness of the walls, underground mining was adopted. While the ore was being removed from the open pit, shafts were in several instances sunk into the foot-wall, the intention being to mine the ore with breast-stopes of an approximate height of 20 ft., followed by underhand stopes of the same height, leaving floors between of the necessary thickness to support the walls, or to effect the removal of the ore by means of what is known as the "stall system." As work progressed, however, it was found that the walls (which are chlorite) were too weak to permit the working of breast-stopes 20 ft. high, there being frequent heavy falls of ground from the hanging-wall, and sometimes from the foot. The plan of following breast-stopes with underhand stopes was, therefore, abandoned. By working breast-stopes only, but little more than one-half of the ore could be removed, and that only at an excessive cost, the ore being so hard that power-drills with $3\frac{1}{8}$-in. pistons, and 6-in. stroke, working under 60 lb. of compressed air, are able to drill but 6 ft. in 10 hours as a yearly average, while from 6 in. to 2 ft. is a common result of 10 hours' drilling. It was plain that some other system of mining must be adopted, and it was proposed to sink by levels of 75 ft., carrying in the cross-cuts from the shafts and working out the ore each way from the shaft from foot to hanging, and from 15 to 20 ft. in height. When this has been done, drift sets consisting of caps and legs are set up the whole length of the opening and connected with the cross-cut; the necessary openings for ladder ways and chutes or mills are timbered from the floor to a few feet above the top of the drift sets; loose rock is run in, and the opening is filled to such a height that from 2 to 5 ft. of loose rock will be over the timber. About 10 ft. of the roof (or back) is then blasted down, broken up, and shoveled into the chutes, from which it is let into the tram-cars standing in the drift.

"As the stope is extended filling follows, rock being let in and the chutes and ladder ways cribbed up as before. It has been the practice to construct these chutes and ladder ways (which measure 5 × 5 ft. inside) of round timber, flattened at the ends, to line the sides of the chutes with plank, placed vertically, and to cover the bottom with short pieces of rails. Each chute is also provided with an iron spout, so adjusted that it may be raised or lowered, for running ore into cars. It was thought that these chutes would wear out before the ore in the roof was exhausted, but, should this occur, a ladder way could be converted into a chute.

"The rock for filling is obtained by putting raises, either in the foot or hanging, close to the ore, from the first level to the open pit, from the second level to the first, and so on. The raises are cribbed through the different levels, and when rock is wanted at one of the upper levels it is obtained by filling the raise below that point with rock or by inserting timber to prevent the rock descending below the place at which it is needed, and where it is run into tram-cars. Two years' trial of this system proved it to

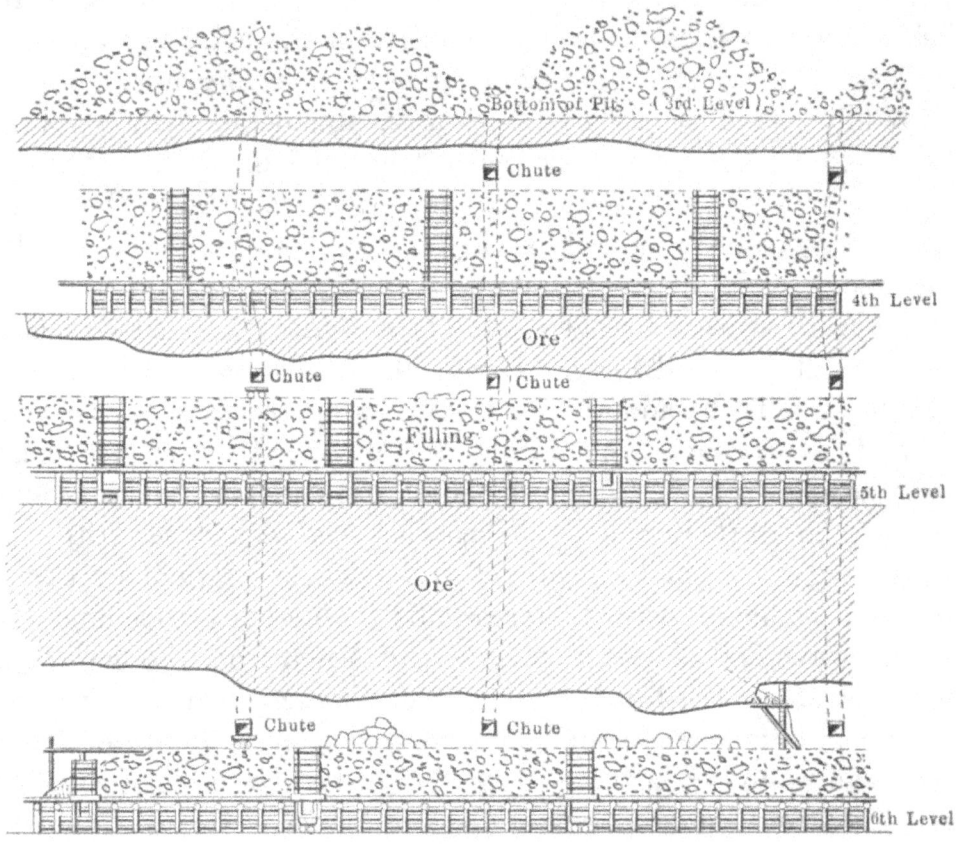

Fig. 72

be satisfactory. In the winter, when the superincumbent rock is frozen, it has been found possible to remove nearly all of the ore in the upper levels, and, after the first level is exhausted, rock is taken from it for those below. Rock to fill the first level, and also any other levels that may need filling before the first level is exhausted, is largely supplied from the accumulation from the loose sides of the pit, and when that is not sufficient, the walls of the open pit are blasted in. In narrow veins the needed filling is sometimes obtained by breaking down the walls underground, and in other cases piles of waste rock on the surface have been utilized. One advantage of this method is, that the rock broken

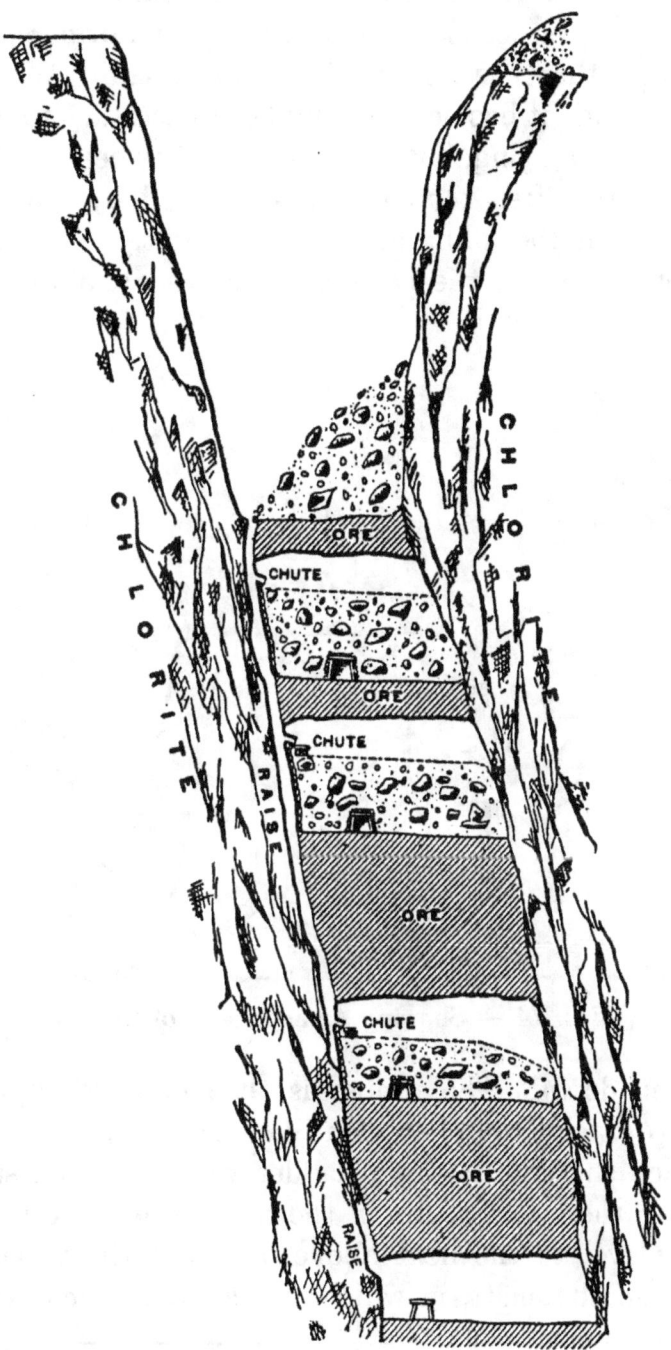

FIG. 73. — Vertical Cross-section, showing the Raise on the Foot-wall

from the walls in blasting, or that coming from seams in the ore, is left almost where it falls, causing no expense for tramming or hoisting and often none for throwing to one side.

"Experience has shown back-stoping to be cheaper than underhand. The roof is always near and easily examined, and as the mills are seldom over 50 ft. apart, the trammers can work at a pile of ore from either side and not delay the drillers, nor stand under ground that has not yet been made safe. When the ore body is very wide it has occasionally been necessary to leave small pillars near the hanging, but the loss of ore has been slight. Slabs in the back are often temporarily supported by cribs of chute-timber, built on the filling and wedged tightly. Holes are then bored over the slabs, the cribs are removed and the dangerous

Fig. 74. — Showing Arrangement of Chutes

rock broken down. The timber is then used to support other slabs, or to extend the chutes and ladder ways.*

"In each shaft is the usual ladder way; but as a sure means of escape in the event of fire in the shaft, raises are put through from each level to the next above, about midway between the ore body and the shaft, in which are ladders which can be used

---

*The small pneumatic hammer drills now in use in many mines are most excellent devices for block-holing dangerous slabs hanging in the stope, as they will drill a hole quickly and with a minimum jarring of the threatening rock, which the heavy strokes of a large machine-drill may cause to fall when not wanted.

should the necessity arise. It has not been found necessary to select particularly large timbers for drift sets or to place them very close together, 80 ft. of broken rock on some of the sets having broken none of the timber."

The accompanying illustrations will make this system of mining plain. Fig. 72 is a longitudinal section, showing the progress of the work on three levels. Fig. 73 is a vertical cross-section, showing the raise on the foot-wall. Fig. 74 is a section showing the arrangement and construction of the chutes.

Chapter XIX

THE CAVING SYSTEM PRACTISED AT THE PEWABIC MINE, IRON MOUNTAIN, MICHIGAN

In some of the iron mines of Michigan and Minnesota, what is known as the caving system of mining is in vogue. This method of removing ore with the use of little or no timber has been described by E. F. Brown in the proceedings of the Lake Superior Mining Institute.

"The system of mining herein described has been in successful operation at the Pewabic Mine, Iron Mountain, Michigan, for a period of three years, and about six hundred thousand tons of iron ore have been extracted from the points in the mine where it has been found advantageous to make use of this method exclusively. This system has been employed most extensively upon a large body of highly silicious, hard hematite ore, of uniform quality. For convenience in operating, the ore body is first divided into blocks, having a length, on the trend of the strata, of about 250 ft., the width of the blocks depending upon the distance between the slate walls, or foot and hanging. The height is the distance between levels, ordinarily 100 ft. The accompanying sketches show one of these blocks of ore in various stages of preparation for mining, also the condition of the over-burden, consisting of broken sandstone and old timber, which had been caved by earlier mining operations.

"The plan of cross-cutting and raising in blocking out the ore is shown in Fig. 75; also the location of the ore with reference to the main level. The main level is driven in the hanging-wall slate, about 20 ft. from the ore body, and parallel with its trend. From the main level, cross-cuts are extended across the ore body to the foot-wall. We usually make four cross-cuts, thus dividing a block of ground 250 ft. long into three blocks of about 80 ft. each. Adjacent to the two cross-cuts, which mark the eastern and western boundaries of the original block, we start raises

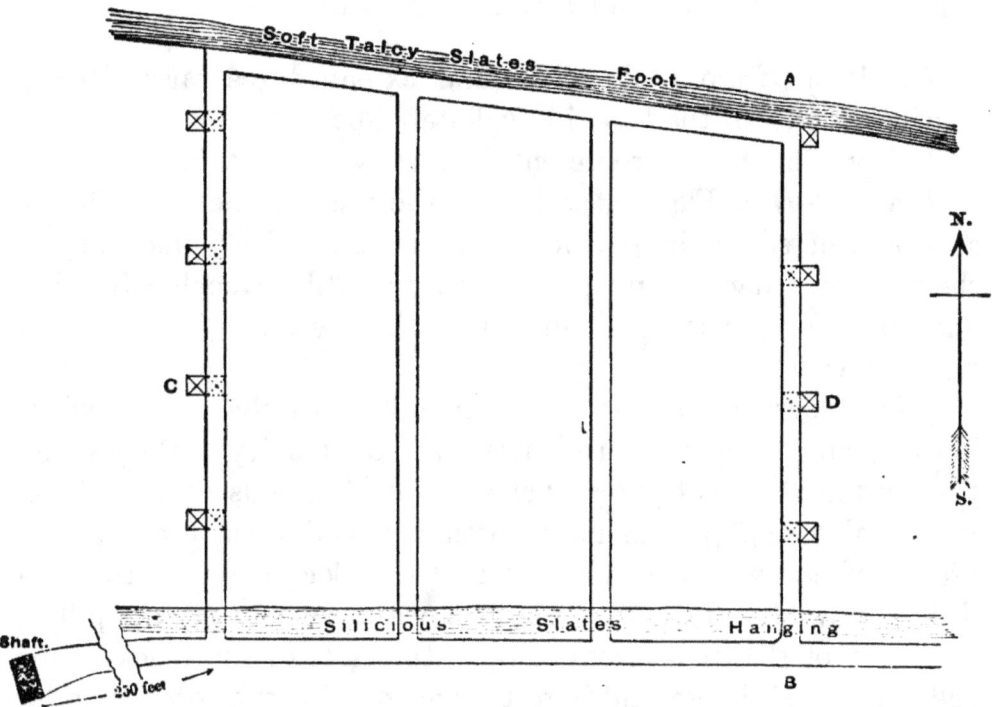

FIG. 75. — Showing Drifts, Cross-cuts and Raises Preparatory to "Block Caving," Pewabic Mine

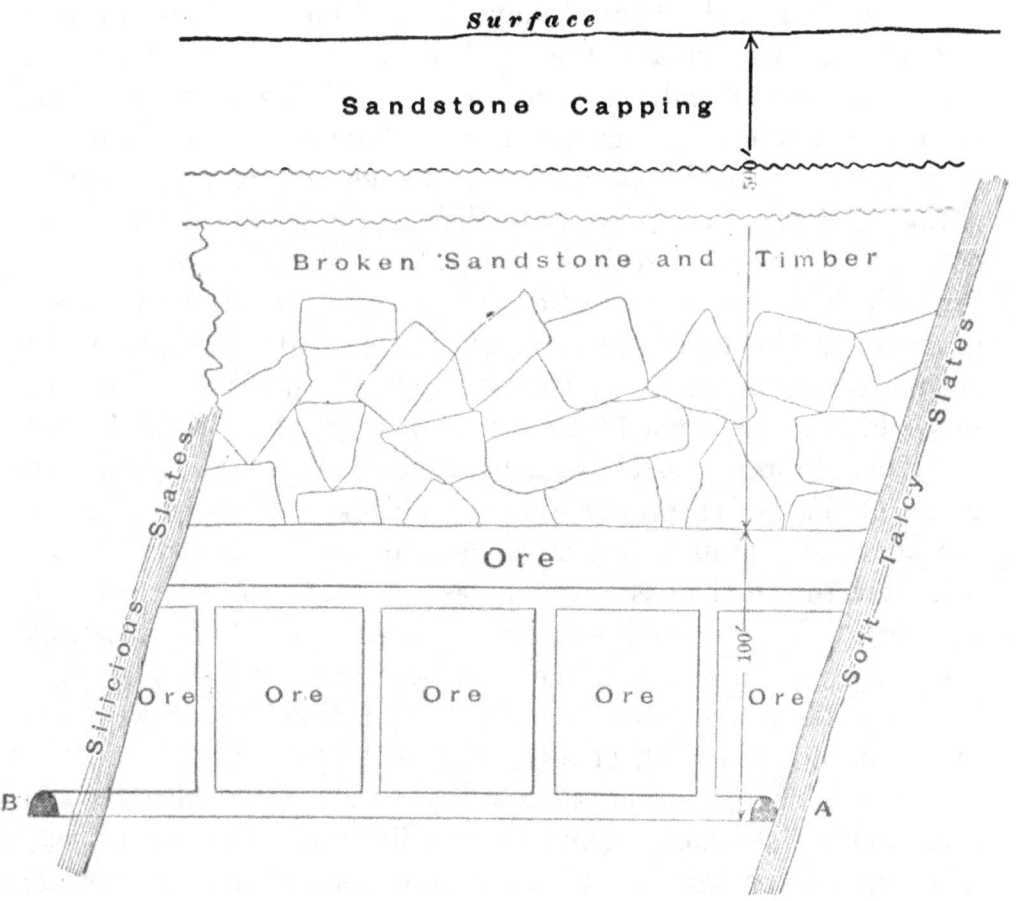

FIG. 76. — Cross-section, "Block Caving," Pewabic Mine

50 ft. distant from each other, and extend these raises 80 ft., or within 20 ft. of the level immediately above.

"Figs. 76 and 77 represent vertical sections taken through A-B and C-D of Fig. 75, and show the raises in detail. Chutes are constructed at the bottom of the raises, and, commencing at the top, underhand stoping is carried on until a trench 8 ft. wide and 80 ft. deep has been cut across the ore body from foot to hanging wall.

"In underhand stoping the ore falls into the chutes when blasted, and is run into the tram cars by gravity. During the time required to cut the trenches across the ends of the block, we are also employed in undercutting it, and when the trenches are completed we have the bottom of the block about in the condition shown by Fig. 78. That is, we have two strong pillars alongside of the main cross-cuts on the ends of the block, while the balance of the ground is resting on small legs of ore, irregular in size and shape. These legs and pillars are then made as small as is consistent with safety to the miners. The remaining portion is then drilled, and when all have been drilled they are blasted in sections. The ends of the block being cut off and the entire block thoroughly undercut, the ore soon begins to cave. This usually occurs in large masses of many hundreds of tons each.

"To illustrate the hardness of the ground and the results obtained, it may be well to state that no timber is used in prosecuting the work just outlined, and that it requires several weeks for a block of this ore to settle a distance of 7 ft., but within six or eight months the ore becomes crushed so that four-fifths of the entire block could be put through a 3-in. opening. After the block of ore has settled down to the original level, and been sufficiently crushed to allow easy driving of cross-cuts, the two cross-cuts in the central portion of the block, as shown in Fig. 79, are advanced through the caved ore, and from these cross-cuts drifts are run to the eastern and western boundaries of the caved territory. These drifts are usually located about 25 ft. apart, on the line of the cross-cut. The drift nearest the foot-wall is connected with the adjacent cross-cut, and from this drift short cross-cuts are extended, at intervals, to the foot-wall.

"These drifts and cross-cuts are all closely timbered, and after the initial timbering require but little attention in the way of repairs. The work of mining is now commenced by allowing the caved ore to run into the drifts and cross-cuts through the opening in the end, or breast, of each. The broken ore is simply shoveled into the tram-cars and taken to the shafts, the breast of the drifts always being full of broken ore. When a full output is desired, the crew consists of one miner and four shovelers at each of the points in operation. Two of the shovelers are always employed in filling the cars, while the other two are tramming to the main level. No drilling is required and but little powder is used in blasting.

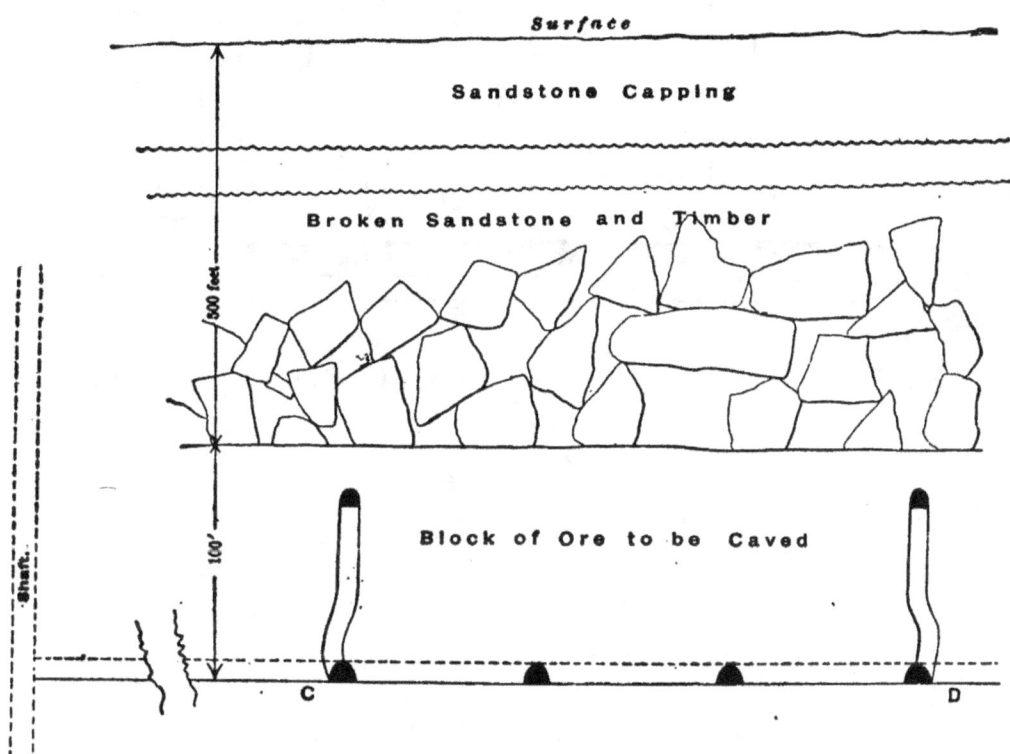

Fig. 77. — Longitudinal Section, "Block Caving," Pewabic Mine

"Drawing the ore in this manner forms funnel-shaped spaces in the body of broken ore, and these are eventually filled by the caving of the timber and broken sandstone from the old level above. When the ore has been so far removed, at any point, that the waste material shows in the working faces, we blast a few of the timbers in the drift and draw our point of operations nearer to one of the main cross-cuts. This operation is continued, and repeated until the ore in the territory formerly traversed by the drifts has been exhausted and replaced by the waste from above. Other drifts from the main cross-cuts are then driven in

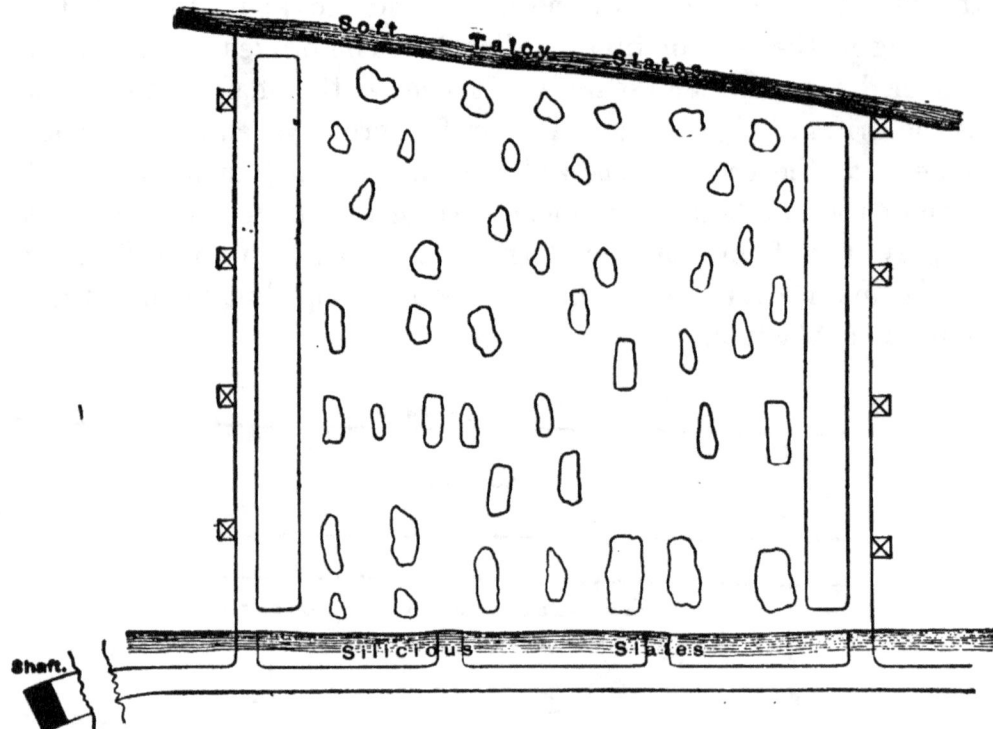

Fig. 78. — Showing Block of Ground ready for Caving, Pewabic Mine

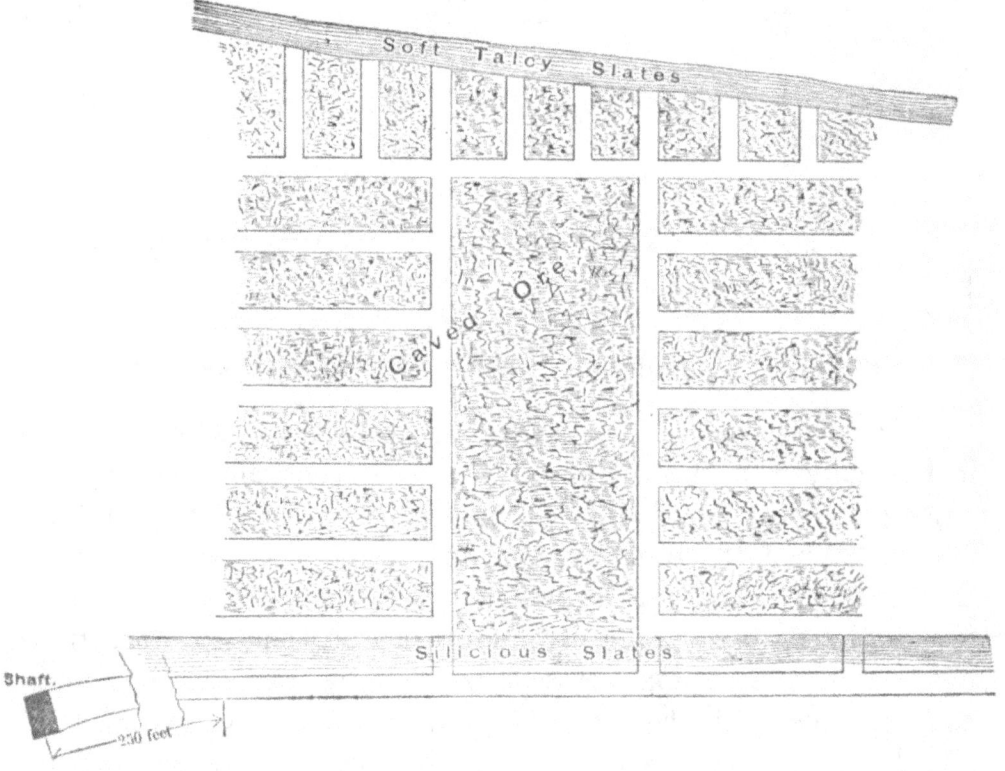

Fig. 79. — Showing Drifts and Cross-cuts through the Block after being Caved, Pewabic Mine

the pillars remaining between the first drifts, and the operation repeated as in the first work. In actual work, it is found possible to draw the ore for three or four feet on either side of these drifts: consequently the second set of drifts practically cleans up the broken ore.

"The typical block of ore herein described contained, originally, about 350,000 tons of ore, and it has yielded about 120,000 tons during the first shipping season. A much larger output could have been attained if desired. We have never had any fatal accidents in connection with operating under this system, and it can, without exaggeration, be called a safe method of mining. With modifications to accommodate varying conditions, this system has been applied to the mining of soft ore bodies, and the results obtained in this direction have been satisfactory."

Chapter XX

STOPING IN SWELLING GROUND

In stoping in veins where filling is necessary, it is not always expedient or possible, without great expense, to obtain the necessary filling from or near the surface. In some of the mines of California, where the veins are wide (as the Utica, at Angels, in Calaveras County), or of moderate width (as at the Gwin mine), the necessary filling is secured by excavating chambers in the country rock of the walls — usually on the hanging side. This method has been found to meet the usual requirements, but it is a procedure to which considerable expense attaches — more, as a rule, than in those methods wherein the filling is obtained at the surface. The mining methods in vogue in California, particularly along the Mother Lode, are mostly the survival of the antiquated practice of early days, many of them most primitive, and, viewed from the standpoint of economy, have little to recommend them. One very noticeable policy in these mines is that of demanding prompt returns from ore development — a procedure often accompanied by expensive and makeshift methods. On the other hand, the peculiar geological conditions obtaining in many of these mines are such that extensive development of ore bodies is unwise, owing to the expense of keeping the workings on the vein open for considerable periods.

In mines having hard-rock walls, as many levels as are desired may be opened, and thousands of tons of ore may be thus advantageously exposed; but where, as in numerous mines of the Mother Lode, heavy, swelling ground is common, it has been found to be the part of wisdom not to open too extensive workings, because of the greatly increased expense of retimbering the levels, mill holes, chutes, and other cuttings. In these latter mines, the usual practice is to drive a cross-cut or drift from the station to the vein. This may or may not at once encounter ore; if not, a drift is driven along the fissure, and this drift must be timbered in the most substantial manner. Ordinarily these drifts are not less than 7 ft. high, 4 to 5 ft. wide at top, and 7 to 8 ft. wide at the bottom. These dimensions are all inside the timbers. Along the fissure it makes little difference whether it be filled with ore or not: this ground is generally heavy. The fissures are usually 10 to 40 ft. between walls, and sometimes greater widths are found — occasionally 100 ft. As soon as ore is encountered, stoping begins and the ore is hoisted and sent to mill; drifting continues, and while overhand stopes are carried up development proceeds. It may be several hundred feet to the limit of the property or of the ore shoot — often over 1000 ft., and the main workings in this heavy, swelling, sometimes running ground, must be kept open until the entire level has been explored, and all the ore between this level and the next above (generally 100 ft.) has been extracted and hoisted to the surface. Some idea of the character of this ground and the expense of working through it by this method may be gained from the statement that in a certain instance a drift of the usual size was run under contract a distance of 200 ft., and headway was made at the rate of 5 ft. daily, but the work could not be completed before it became necessary to return to the portion first driven for the purpose of retimbering. The timbers generally employed in working ground of this kind are from 20 to 30 in. in diameter, but it is by no means uncommon to see these big logs, split, crushed, and broken within a few weeks of the time they were placed in position underground.

Main Gangways Driven in the Country Rock

The advisability of cutting the main gangways in the hard rock of the walls, either foot or hanging, and developing and exploiting the vein through a series of cross-cuts, is emphasized. These main gangways, being driven in, say, the foot-wall, should have cross-cuts run at right angles to the main gangway. These cross-cuts should be disposed at regular intervals for the purpose of prospecting the fissure and rendering the ore discovered easily accessible. The distance of the main gangway from the vein must always be determined by the character of the rock in which it is driven, and varies from 20 to 60 ft. In all cases it must be sufficient to avoid the zone of rock, which will swell upon exposure. The intervals between these cross-cuts must be determined by the character of the ground adjacent to the vein and by the vein material itself. When the ground is very bad, the cross-cuts must be closer than where the ground is more favorable and less likely to crush the timbers. Ordinarily, if run at intervals of 240 ft., the distance will be found convenient. Raises should always be put through to the level above before stoping is commenced. Too often this important matter is neglected, owing, as already stated, to the desire to realize a profit on the ore as quickly as possible without indulging in anything that savors (under the mistaken idea of economy) of unnecessary expense. The raises may be driven either at the ends or the middle of each stope section. They should be divided into two compartments, and cut about 15 ft. from each cross-cut. This would leave an intervening vein section of 210 ft., in which distance mill holes could be carried up at intervals of 30 ft. as stoping progresses upward. Generally speaking, when cutting a raise it is good practice to sink a winze immediately over it from the level above, to make a connection with as little delay as possible, in order that good ventilation may be secured and an additional exit for the miners provided. The increased efficiency of the workmen will quickly pay for the added cost of sinking the winze. The timber used in the stope, if any be required, may also be lowered through one compartment of the raise, which is far better than hoisting it and machines as well from the level below, though the use of small hoists, operated by independent electric motors, has greatly simplified the work of taking machines, timbers and other heavy articles up into stopes.

This recalls a home-made device introduced into the South Eureka mine, near Sutter Creek, California, where, by an arrangement of three grooved sheaves — two at the top of the raise and one at the bottom — timbers were hauled up into the stope with comparative ease. One of the upper sheaves was set over one compartment of the raise, the other over the second compartment. The third wheel was fixed at the foot of the raise. The rope passed up the raise over the two upper sheaves, then downward and under the third sheave at the foot of the raise. When it was desired to take timber into the stope, the empty bucket was pulled by windlass up to the upper part of the raise, and the timber was made fast, down on the level, to the rope in the opposite compartment. The bucket was then filled with ore, and its weight, acting as a counterbalance, made the hoisting of the timber a matter of small difficulty. The scheme was rather dangerous in any hands but those of cool men who, with a proper code of signals, worked the device in a satisfactory manner.

Returning to our stopes, when the lateral drift and cross-cuts have been completed, and drifts run in the vein on the level of the foot-wall drift, the raise put through, chutes built, and all preliminary arrangements completed, stoping may begin at as many places as there are raises and mill holes started. As soon as connection has been made from one section of the stope to the next, ventilation should be as good as it is possible to make it, this depending, of course, upon the raises having been put through to the level above. The ore should now be removed rapidly with the use of no more timbers than necessary. When the ore has been taken out, if the excavation is not filled, the walls will soon close in and the stope be lost forever. This generally should result in no particular harm, as the main lateral drift in the foot-wall remains open. Where the squeezing of the ground, following the removal of ore, interferes with the further progress of stoping, filling may be obtained from the walls by cutting chambers, as shown in Fig. 80. There will be observed two chambers, one inclined upward from the plane of the hanging-wall side of the vein; the other running in flat. The first is possible where the wall is firm and there is little danger of the rocks overhead falling. The second is used where the rock of the hanging-wall, immediately adjacent to the vein, is loose and likely to cave. In such cases a cross-cut is run directly into the hanging, and, if need be, the ground may be timbered until firm rock is reached beyond, where the chamber may be cut, either flat as shown, or carried upward at an angle. It is evident that the lower chamber shown in Fig. 80 has the advantage, as all rock broken therein will pass out and downward into the stope beneath by gravity, making shoveling unnecessary. This method of stoping and filling in deep levels will be found a great improvement on some of the old and time-honored practices on the Mother Lode of California.

In some mines, if stoping be expeditiously carried on in the manner here suggested, no filling will be required, the timbers affording all the support to the walls that may be required; but in most mines filling is indispensable. The chambers cut to supply filling may be mined either by hand or with machines. By means of a lateral foot-wall drift and cross-cuts a large section of the vein may be completely mined in a few months, whereas by the

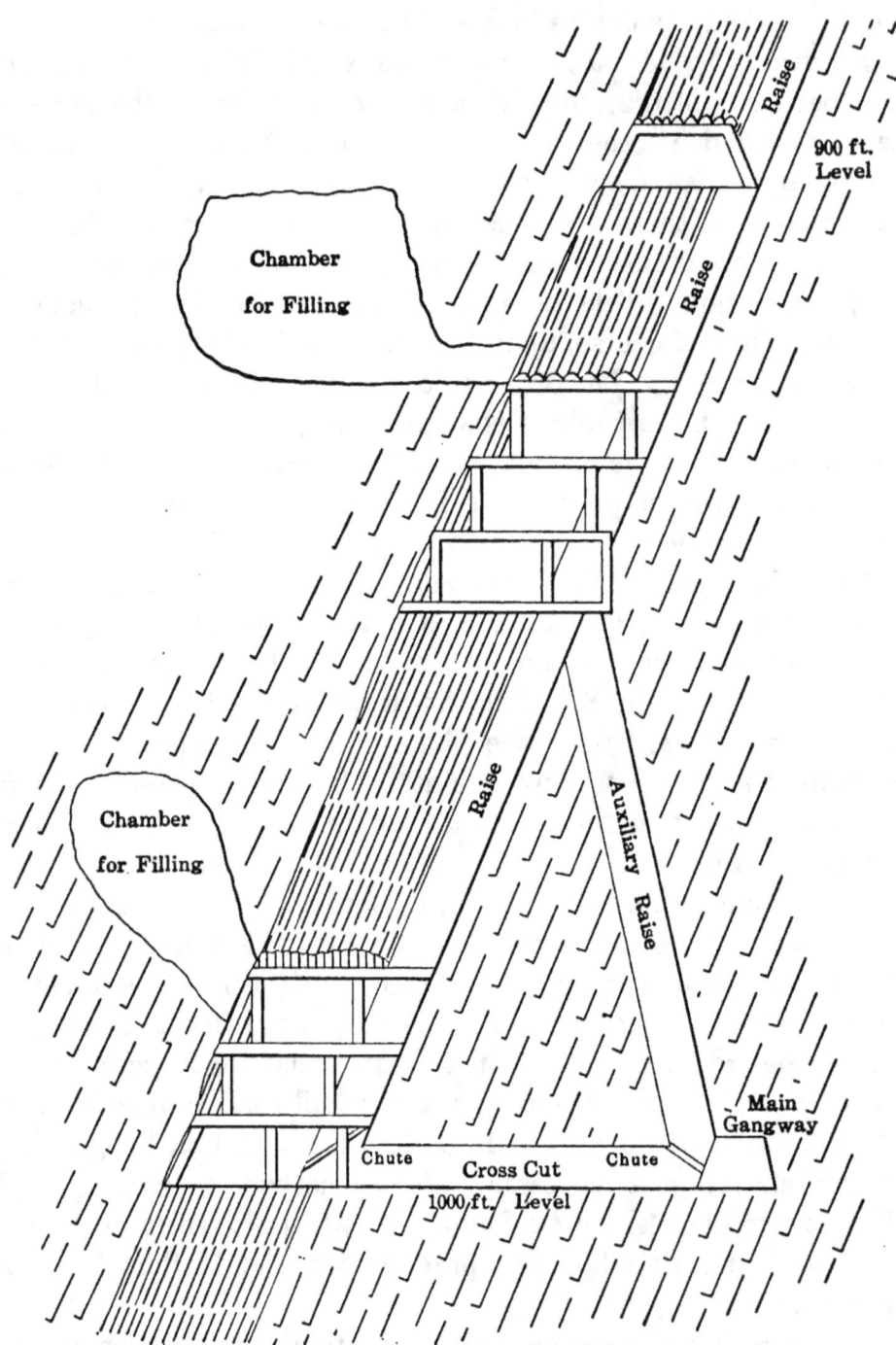

Fig. 80. — Proposed Method of Mining and Filling from the Walls.

old methods the main workings must be kept open for one, two, or three years, or even a greater length of time. A further advantage, consisting in greater expedition, may be gained by cutting an inclined raise from the main gangway in the foot-wall to a point about midway between two main levels, and there opening an intermediate level (see Fig. 80). This scheme is advisable where the ground is particularly bad, requiring much reinforcement of timbers and renewals as well, if the stope is to be kept open very long. It is also a good method for lessees who wish to remove as much ore as possible within a given period. By this method the ground need be kept open only about half as long as would otherwise be necessary if the stope were carried up from

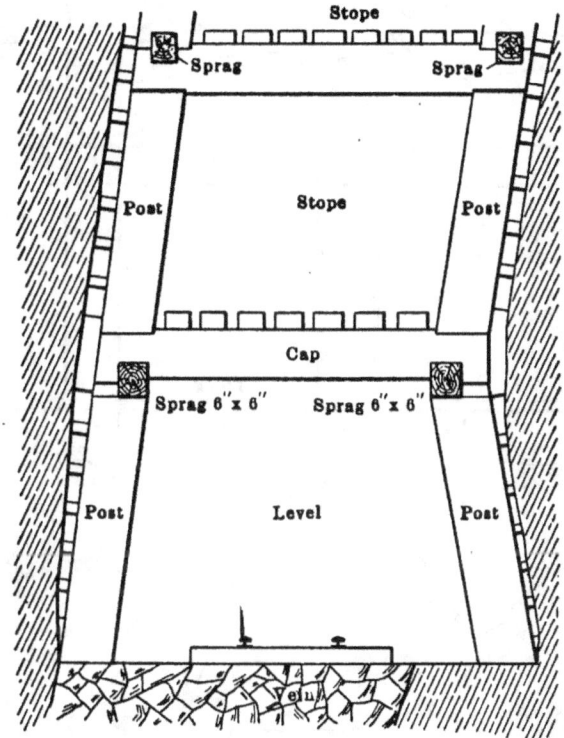

Fig. 81. — Timbering in Stopes at the Gwin Mine

the main level only. The additional expense would be in cutting a series of inclined raises, each about 45 to 50 ft. long, depending somewhat on the dip of the vein. The working out of the details in each particular case must be left to the superintendent, as the conditions at no two mines are exactly identical. The method proposed appears in Fig. 80, though doubtless subject to modifications and possibly to improvements. Fig. 81 shows the novel method of timbering and stoping practised at the Gwin mine, where the walls were often so insecure in places as to require the unusual method of timbering shown.

In sinking winzes, it is an excellent idea to make arrangements to carry the track several feet higher than the floor of the level, so that the bucket or skip may be dumped automatically into a car or a bin. In square-sets the track can be continued up into the second set, where a bin could be built provided with a simple chute for drawing the rock into cars.

Connecting Levels When Stoping Veins of Moderate Width

When stoping by the overhand method, on approaching the

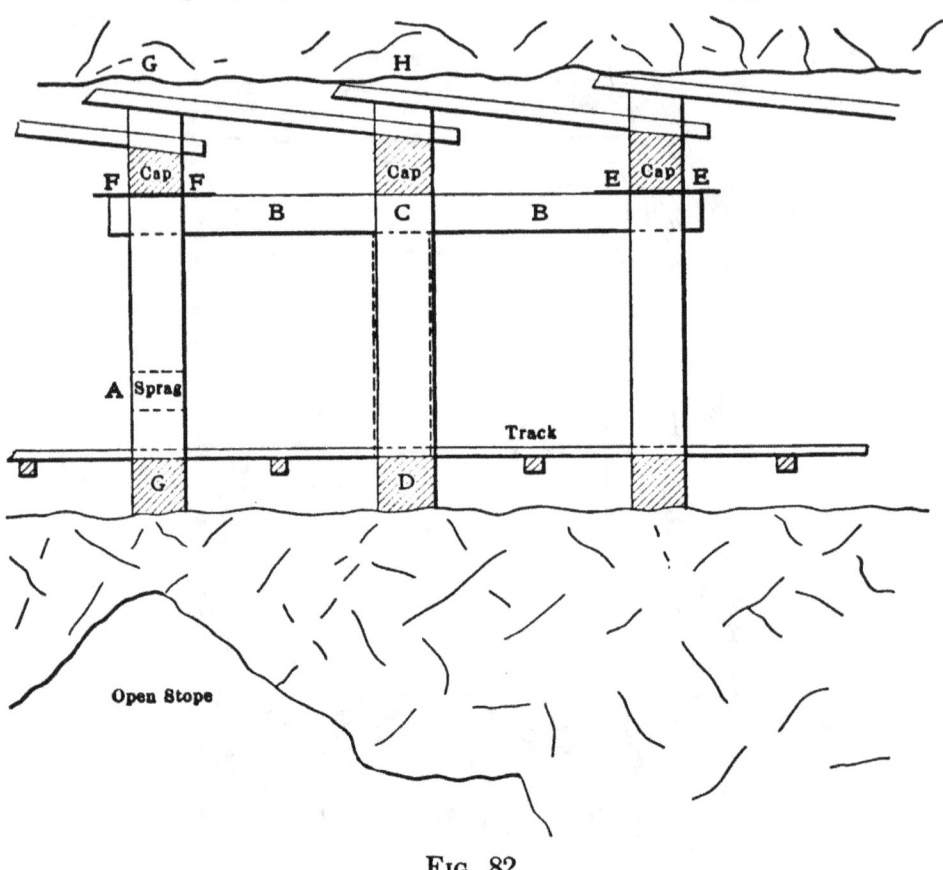

Fig. 82

floor of the level above it is necessary, where posts and caps (the usual drift set) have been used, whether sills were employed or not, to take some precaution to prevent the falling out of those timbers and the caving of the filled stope above. At the Bi-Metallic mine, near Phillipsburg, Montana, an expeditious and safe method was introduced (see Fig. 82). When ready to break through the floor under any particular set of timbers in the gangway above, a sprag was placed between the posts, a few inches above the floor, at A, and there tightly secured with shingle wedges. A heavy stick of timber, *B B*, long enough to reach across three sets (kept on each level of the mine for this purpose),

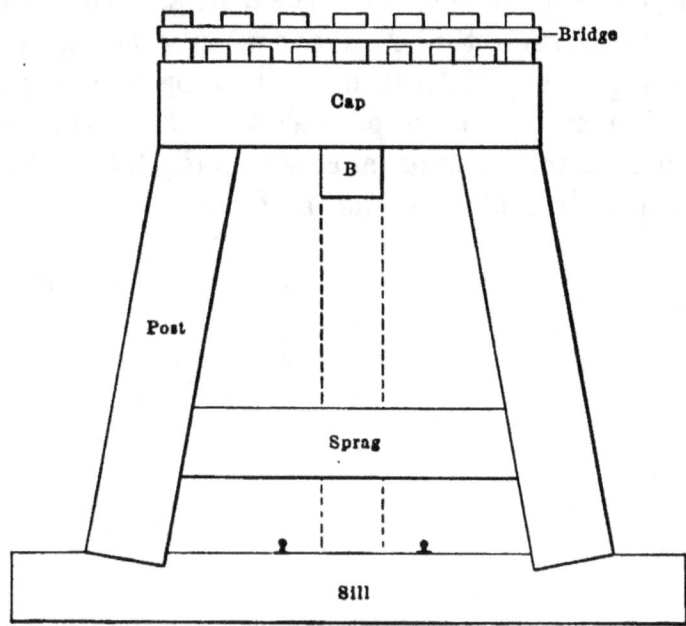

Fig. 83. — Section at G.G.

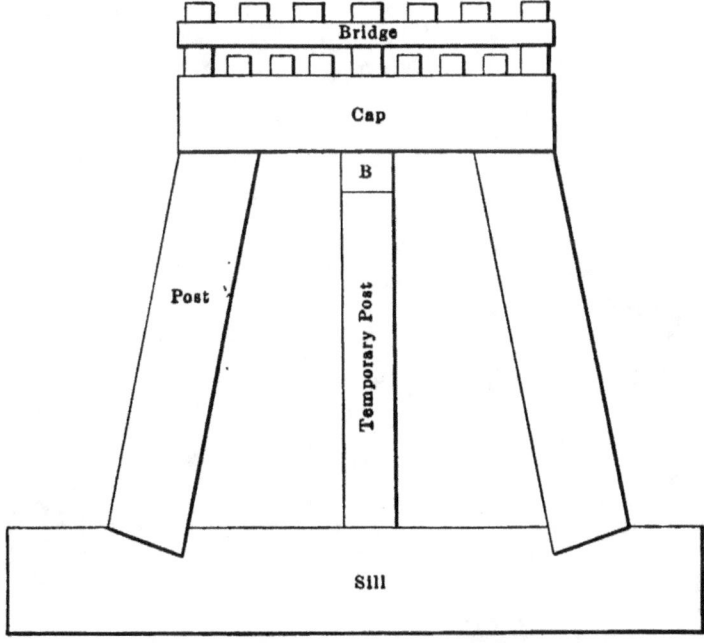

Fig. 84 — Section at D.H.

was lifted to the roof of the gangway, midway between the posts, one end being beneath the cap forming a part of the set about to be undermined. This timber acts as a lever, having a fulcrum at C in the post which supports it, the foot resting in the center of the drift on a sill at D. Wedges were driven in tightly at E, and also at F, when the bottom of the drift at G could be removed with safety to the set of timbers above it, only the sill (if there were one) dropping out. The set may then be connected securely with the timbers in the stope below. This excellent scheme may be modified to suit unusual cases. Figs. 83 and 84 illustrate sections in the drift at GG and DH.

Chapter XXI

STOPING LARGE ORE BODIES

In previous chapters have been described at considerable length the various methods ordinarily employed in mining veins and ore bodies of moderate width, and the various methods described are those principally used throughout the world. Besides these, however, there are the special methods required for particular cases, where the low dip of the vein, or some other physical condition, requires such methods. Prior to 1860, veins under 20 ft. in width that required artificial support of some kind were rarely mined extensively, though many attempts were made to do so, usually attended with disastrous results and generally loss of life, as well as of thousands of tons of ore. In these attempts some ingenious schemes were introduced, including splicing of long timbers, which were supported at intervals by braces or sprags, these being placed longitudinally with the vein. Notwithstanding the ingenuity of the miners and the excellence and daring of the schemes, caves followed almost every attempt. Moreover, the expense of such expedients was in most instances prohibitive. When the mines of the Comstock Lode were first opened, ore deposits of unusual width were found. Had these been low grade and had they offered no unusual inducement, it is not unlikely that the tremendous advances in the art of mining that then came into existence would have been delayed for many years, but the ore deposits of the Comstock were as phenomenal in richness as they were in size.

The following interesting reference to early mining on the Comstock is from Monograph IV of the United States Geological Survey, "Comstock Mining and Miners," by Elliott Lord: "At the 50-ft. level [of the Ophir Mine] the vein of black sulphurets was only three or four inches thick, and could readily be extracted through a drift along its line, propping up the walls and roof, when necessary, by simple uprights and caps. As the ledge descended the sulphuret vein grew broader, until at a depth of 175 ft. it was 65 ft. in width, and the miners were at a loss how to proceed, for the ore was so soft and crumbling that pillars could not be left to support the roof. They spliced timber together to hold up the caving ground, but these jointed props were too weak and illy supported to withstand the pressure upon them, and were constantly broken and thrown out of place. The situation was a curious one. Surrounded by riches, they were unable to carry them off. The company was at a loss what to do, but finally

FIG. 84 a. — Remarkable Stope Timbering 1250 Level of the Elkhorn Mine, Montana. The Roof consists of Slate which is Difficult to Hold.

secured the services of Philip Deidesheimer, of Georgetown, California, who visited and inspected the treasures of the Ophir."

Introduction of the Square-Set by Philip Deidesheimer on the Comstock

That the ore body could not be extracted in the usual manner was at once apparent, and Mr. Deidesheimer told the writer of this treatise that he set about his task with some misgivings. He did not at one stride grasp the idea of the square-set, but the system which now bears his name was the outgrowth of circumstances and the very necessities of the case. He instituted a policy, however, the wisdom of which soon became apparent. The first step was to cross-cut the vein from wall to wall, starting from a drift on the hanging-wall side of the vein. As the work advanced he set up posts and placed caps above them — not across the course of the drift, as usually done, but along the sides, the idea being to form, when completed, a line of caps that would reach continuously from wall to wall. To accomplish this the ends of two caps were placed upon each post, except at the ends. These novel sets were held in place by pieces of 2 × 4-in. scantling 4½ ft. in length and reaching across the drift from near the top of a post to that opposite.

Having successfully driven the cross-cut Mr. Deidesheimer now ran a drift some distance along the foot-wall, timbering with posts and caps in the ordinary manner; that is, the caps were placed upon the posts at right angles to the drift and parallel with those in the cross-cut. The posts in each case were set vertically. Returning to the point where these operations were begun, a second section by the side of the first cross-cut was taken out and timbered with a single line of posts and caps, the 2 × 4 scantling being placed as in the first case. When this section was completed there were standing three rows of posts surmounted by three lines of caps, extending from the foot to the hanging-wall. This was not really a new idea, as Mr. Deidesheimer had previously employed the same method in his draft mine on Forest Hill, California, where the breast was carried 125 ft. wide, the roof being supported by rows of posts with continuous caps. The work thus far performed in the Ophir revealed the fact that an extremely rich body of ore extended upward from the level where this work had been done. The miners were directed to commence stoping upward in the body of soft, black, crumbling ore; soon a considerable excavation had been made, and it also became evident that the ground must be secured at once.

In the Georgetown mine, which Mr. Deidesheimer had left but a short time before, the vein was vertical, and the walls were so soft and crumbling that, in order to safely stope out the mineral, he had resorted to the expedient of setting one post directly above another, the lower end resting on the cap, and in this way he managed to work the vein without much difficulty. (Compare

Fig. 81.) The Georgetown experience suggested the idea of adopting a similar plan in the Ophir. Accordingly, Mr. Deidesheimer had a mortise cut at the junction of two caps, which were already in place, and, having a post framed with a tenon to fit, set the post directly above the one resting on the floor below. In a short time four posts were in position with the caps upon them as below, together with the frail 2 × 4 scantling, the office of which was to keep the other timbers from falling down. The first square-set timbers, it will be seen, were framed in the mine, the mortises being cut in the timbers in place. The work of extracting ore proceeded slowly yet, for the ground had to be secured as well as possible. It soon became evident, however, that something more substantial than 2 × 4 scantling would be required to keep the timbers in position, and it was determined to put in timbers of the same dimensions as those forming caps and posts. This was done at once, and the square-set was complete in principle, though not in detail. The caps occupied all the space on top of the posts, leaving no resting-place for the ties, which had to be supported in some other manner. As they were looked upon as simply an auxiliary — a support to the posts and caps, not subjected to vertical pressure — they were only required to be held in position. Accordingly, a quantity of iron spikes were made, in shape somewhat like the thumb, having a sharp point at one end, the other end having a face three-fourths of an inch square. Two of these spikes were driven into a post at the proper height, and two in the post opposite, the ends projecting, and the tie placed so that the ends rested upon these iron lugs, wedges being driven in at the ends to secure firmness. The posts and caps were now framed on the surface and delivered below, ready for use whenever needed. The work of mining now progressed much more rapidly and the problem seemed solved. Soon after, it was determined to frame the timbers so that the ties might also rest on the posts; and the stopes becoming of great size, the dimensions of the timbers were increased.

As the work progressed, slight changes were made from time to time, whenever any improvement seemed possible. Sills were laid on the floors of the levels as a foundation for the timbers above, which had now assumed massive proportions. The sill timbers were as long as it was possible to get into the mine. The men who were obliged to handle these ponderous timbers could see no reason why the sills should be longer than the caps, and had from the first looked upon the growth of this new system with much prejudice. When the great stopes were carried up from level to level and connected, the wisdom of the use of long sills became apparent, as they permitted the removal of all the ore and the placing of timbers without danger or loss, which could not have been accomplished with short sills, as, when breaking up through the floor of a level from below, short sills would have had nothing to sustain them, and their use would have greatly increased the danger. When, in the course of ore extraction, the work reached the walls, additional timbers, called "wall plates," were put in. The caps were extended from the nearest post to the wall plate, except when a post came within two feet of the wall plate. In such a case the cap extended from the wall plate to the second post in a single piece.

Usefulness of the Square-Set in Extracting large Ore Bodies

Timbering mine excavations with square-sets has become a world-wide practice since the days of 1860, when Philip Deidesheimer introduced the method in the Ophir mine. Prior to that time there was no method known to miners whereby large ore bodies might be safely mined and the walls and roofs of stopes supported. Mr. Deidesheimer soon recognized the weak points in his system and corrected them as fast as they developed. He spared no expense to make his system perfect, or as nearly so as possible, and it is safe to say that the Ophir mine was, early in the sixties, the most elaborately timbered mine in the world. The Ophir timbering omitted nothing calculated to make a complete and enduring system of support. It comprised sills on the floor of each level, posts, caps and ties. These timbers are common to all complete square-set systems, but the Ophir system included, in addition, wall plates and angle-braces, which were inserted diagonally between the sets as shown in the illustrations (Figs. 85 and 86), and were for the purpose of strengthening the system and affording a more direct means of resisting the thrust due to the downward weight of the hanging-wall, as it is impracticable to employ the square-set system in any other manner than by setting the posts upright and the caps and girts horizontally. Of course, the posts will receive any pressure coming vertically from above, but owing to the usual slope of the hanging-wall,

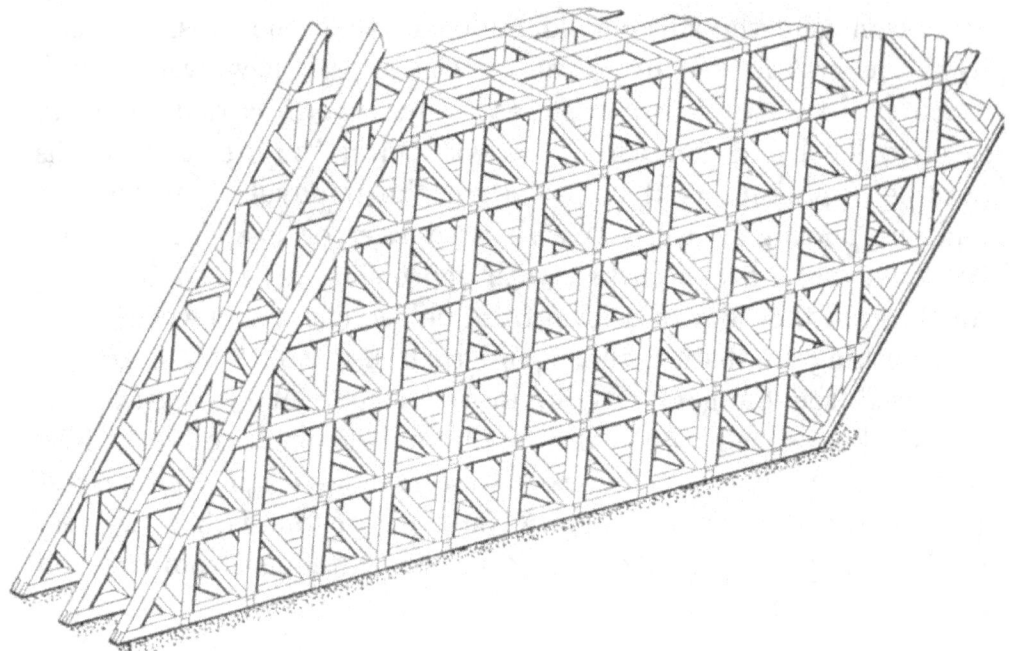

Fig. 85

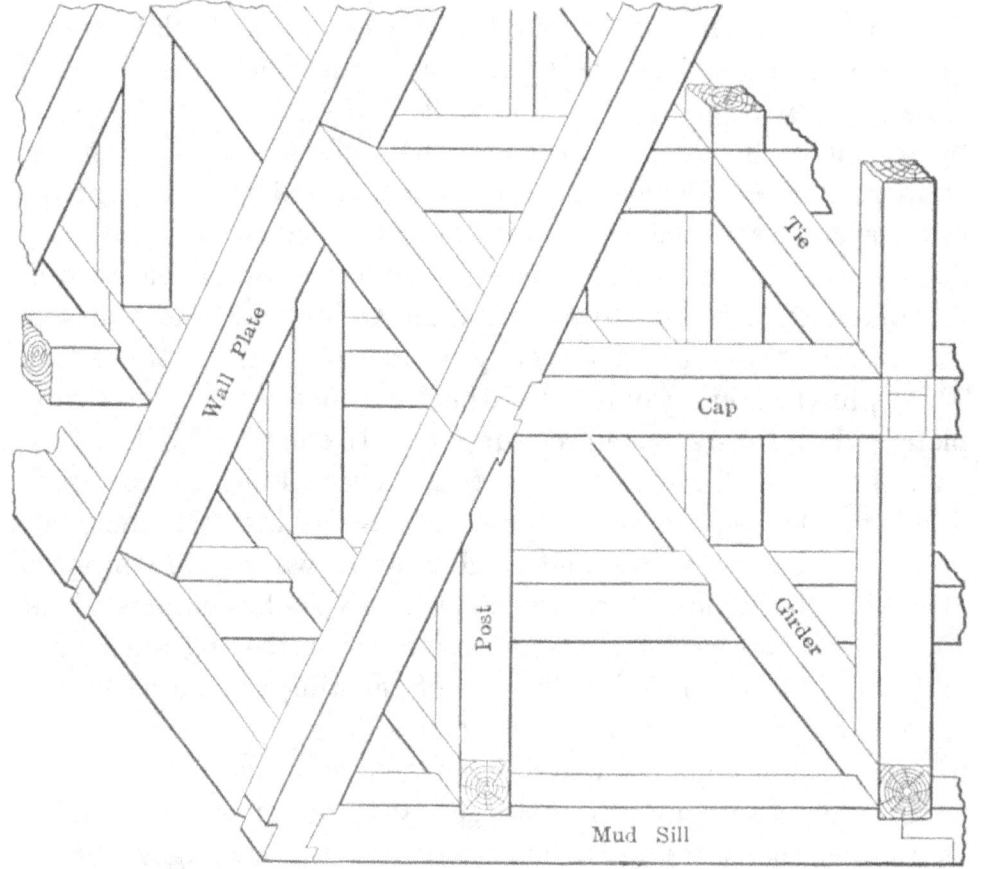

Fig. 86. — Details of Timbering in the Ophir Mine, Comstock Lode

at an angle to the horizon, it is impossible for either posts or caps to directly take this diagonal thrust, due to the weight or settling of the hanging-wall; consequently the angle-braces were introduced into the system. Theoretically, the angle at which the braces are placed must be determined by the angle of slope of the hanging-wall, and this, in turn, would determine the relative height of posts to distance between them. As the angle of dip of any vein wall is seldom constant over large areas, it at once becomes evident that uniformity in the dimensions of timber sets would sooner or later be disturbed by a change in dip in the walls, this necessitating a change in the dimensions of the system. When such a change is made, the symmetry of the system is at once destroyed, for the sets will no longer meet in making connections from one stope to another and the uniformity so necessary to the success of the system is lost.

Early in the history of square-setting these facts were observed and the clever miners of those earlier days endeavored to meet the exigencies of the case by increasing the size of timbers and reinforcing the sets by placing angle-braces, and often by adding auxiliary posts or caps to the sets, placing them at the side of the original posts and beneath and above the original caps. This method, when introduced, indicated a desperate situation, but could be, at best, only a makeshift, for ground so heavy that such elaborate reinforcement of the timber sets became necessary could not be held by any system of timbering short of solidly filling the stope with timbers, which expedient has actually been resorted to in a few cases. Cribs of solid timber, and cribs filled with waste rock, early became features of the square-set system.

Mistakes Made in Using the Square-Set

From the beginning, however, the worth of the square-set system was everywhere recognized, and it was promptly adopted throughout the West. Comstock miners went abroad and introduced the system in other lands. W. H. Patton, a Comstock miner, took the idea with him to Australia and first introduced it there in the Broken Hill mines of New South Wales. Mining men everywhere adopted it, but at the same time they considered Mr. Deidesheimer's system unnecessarily elaborate and expensive and began to dispense with various portions of it. The wall plates were the first to be omitted, then, in many places, it was found the angle-braces could be left out. In some cases the sills were omitted, and still the great system, reduced as it was, admitted of the removal of large blocks of ore with apparent safety. It was not long, however, before the limitations of safety were stretched to the utmost, and usually exceeded. Enormous stopes were excavated in great veins, and the holes were filled with an elaborate system of timbering, comprising posts, caps and ties. In some instances the sets were not properly spragged to the walls or to the roof, while in nearly every instance where the square-set method was introduced, the grave mistake was made of attempting to remove too large a superficial area of ground at one time. The natural result followed; disastrous caves occurred in nearly every great mine where the square system was employed.

It became evident that something more was required to perfect the method. It is a matter worth noting that the great stope in the Ophir mine, on the Comstock, excavated and timbered under the personal direction of Philip Deidesheimer, did not cave. His system, considered by many to be unnecessarily elaborate, at least held the ground until he had excavated a stope 400 ft. high, from 60 to over 100 ft. wide and several hundred feet in length. This was an immense open stope, probably the largest in the world at the time. It was filled after reaching the dimensions indicated, but it is highly improbable that this immense hole could have been made and kept open as long as it was had any part of the original Deidesheimer system been omitted. Had either wall plates or angle-braces or sills been left out, it is believed the Ophir stope would have met the disastrous fate of so many others. The secret lay in carefully preventing the ground from starting. Once the ground gets a start, and it begins to "work," the difficulties of holding it are greatly increased, and it is rarely that the placing of additional timbers in the stope will avert the threatened disaster. We have knowledge of a number of cases where this was attempted, and in each instance the cave occurred, the timber supports being entirely unequal to the emergency of holding up great masses of heavy ground.

The square-set system is sound as originally designed; it is the various modifications of the complete system that are at fault. The omission of important features is disastrous to the general scheme. However, it was long since learned by expensive experience that heavy ground cannot be held by any system of timbering if the excavation be of large size. There are numerous instances which seem to contradict this statement, where large stopes are held by a rather light system of square-sets, but careful examination will, in these cases, generally show that the ground would stand nearly as well with no timber support at all. The Yellow Aster mine at Randsburg, California, furnished excellent examples of this condition, for in that property are stopes several hundred feet in length, over 100 ft. wide, and more than 100 ft. high, timbered with square-sets in which the largest timbers are 10 by 10 in. If the hanging country were to settle on these frail supports they would not resist the pressure an hour. In fact, several caves have occurred in this mine where the timber sets were subjected to pressure. Ground stands well in all desert countries, the dry atmosphere greatly reducing the tendency to shift and cave. In the Calico mining district of California are huge stopes which have been open for twenty years, and there is not a stick of timber in them. The rock is rhyolite-tuff and tuffaceous breccia, and while it is easily mined it rarely caves, though, of course, there is a limit to the extent to which even that excellent standing ground may with safety be excavated without support. Experience has taught that the square-set system of timbering may be applied to the excavation of large masses of ore, but that the excavations must be restricted in superficial area, that the stopes after timbering must be filled as the work progresses, constantly keeping the size of the excavation within the limitations of safety.

In a mine at Angels, Calaveras County, California, a cave occurred several years ago. Report said that the stope, which was about 100 ft. square, had been carefully timbered with massive square-sets. The writer had an opportunity to visit this mine soon after the cave and went through the portions of the stope which were still accessible. The great timbers, 24 to 30 in. in diameter, had been split and broken in every direction, showing the great weight of the rock that crushed it. The superintendent said that "not only was the stope timbered, but it was filled as well." This statement was so incompatible with the fact of the disastrous cave that had occurred, that it led to an investigation of the condition of things. It was found that the stope covered an area of nearly 10,000 square feet — about 100 feet square. It had been carried up a little over four sets high — the sets were

8 ft. high and the back was from 2 to 6 ft. above the upper sets. The hanging-wall dipped at an angle of about 75 to 80°, the foot-wall being somewhat flatter, but consisting of several feet of talc schist — a most excellent lubricant. Filling had been placed in the stope to the height of about 20 to 25 ft. This brought the fill to within 12 to 15 ft. of the back, but in no place did it touch it. Props had been put in between the top of the sets and the back, and also reaching from the sets to the walls. The great mass of ore, being undercut over a large area, began to settle, the process being aided by the talcose walls, and a cave resulted as a natural consequence.

Filling is all-important, but it must be placed where it will afford support to the overhanging rock, or it fails to accomplish the purpose for which it is introduced in the stope. By a judicious combination of square-set timbering and filling, ore deposits of any size and character may be safely extracted.

Chapter XXII

FRAMING SQUARE-SET TIMBERS

The framing of mine timbers is an operation requiring merely ordinary skill on the part of the workman. Measurements must be made with care, and always from the same side of the stick of timber, for the reason that large timbers of stated dimensions often vary from $\frac{1}{4}$ to $\frac{1}{2}$ in. in a nominal length of 20 ft., and sometimes the discrepancies are even greater. It is obvious that a stick of timber 20 ft. long which is 12×12 in. at one end and $12\frac{1}{2} \times 12\frac{3}{4}$ in. at the other, must be framed with due regard to this irregularity in cross-section if the work is to be placed uniform, for if this care is not taken it will quickly result in bringing corners out of line. Ever since mining began and timbers were used for the support of underground workings, all timber framing was done by hand, until a few years ago, when a device consisting of a gang of saws was introduced to frame timbers by machine. These timber-framing machines, simple at first, have been improved until to-day they have reached a high degree of perfection. Timbers are now run up to the machine on a truck and in a moment both ends are completely and accurately framed. At any large mine it is an economy to use a timber-framing machine, as these devices perform the work as well or better than it is generally done by hand, and in one-hundredth part of the time.

The framing of square-set timbers is all done with the same object in view — the removal of small sections from the timbers in such a manner as to form tenons at the ends of the individual sticks. The several pieces of timber making up the system are each framed differently, yet each individual piece of the same kind is framed in exactly the same way, and wherever wanted in the mine one of these pieces will fit in without alteration. A post is a post, and a cap is a cap, and if a post or a cap is wanted in any part of the mine there should be but one kind of post or cap to send to the workmen at the place where it is wanted. This is one of the chief advantages of the Nevada, or Deidesheimer, square-set system. The several individual pieces going to make

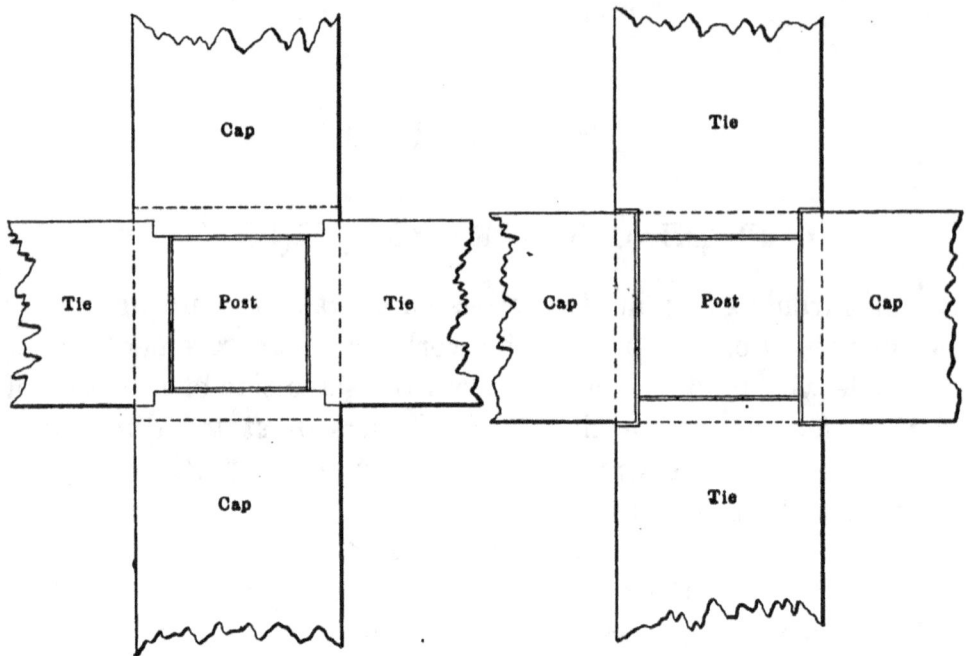

Fig. 87. — Method of Framing for Top Pressure

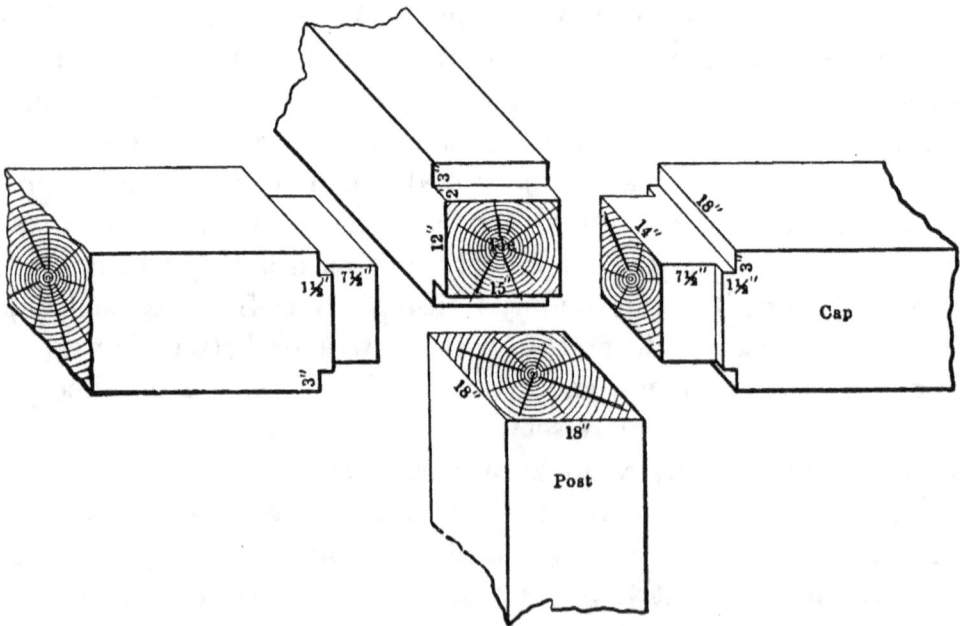

Fig. 88. — Framing to Resist Side Pressure

up the complete square-set system are sills, streak-sills, posts, caps, ties (sometimes called girts), wall plates, angle-braces, butt-caps, cap-sills, extension caps, sprags and — by no means the least important — wedges, for without the simple wedge the entire system fails, and cannot be made to answer any useful purpose.

Figs. 87, 88, and 89 show some of the most important methods of framing square-set timbers. In each case where "square timber" is employed, the post is actually square in section, though the cap and tie may vary from this. Very frequently, where the post and cap are of the same dimensions, the tie is 2 in. narrower than the cap, though of the same depth. The object of framing the timbers is to permit both cap and tie to find a secure

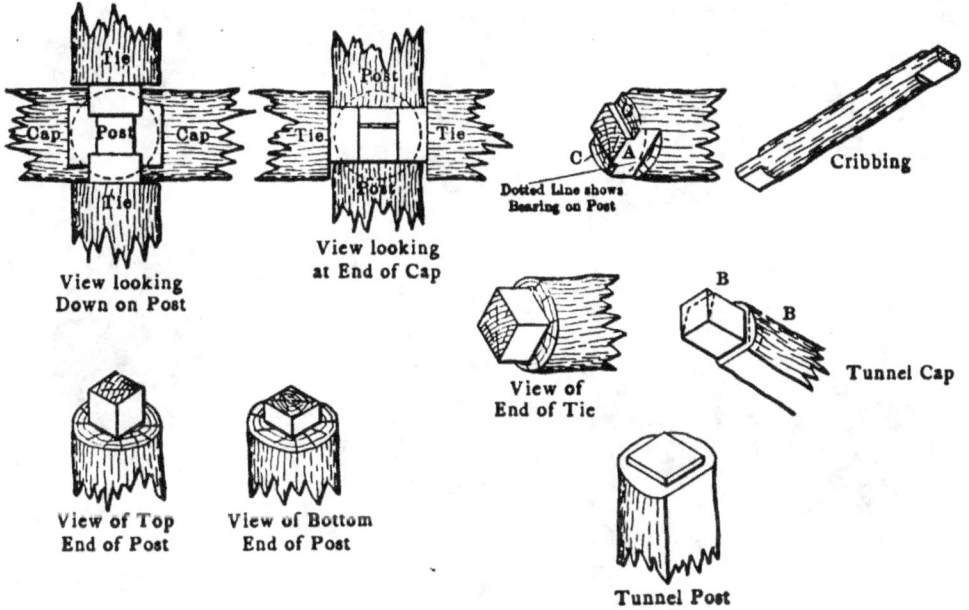

FIG. 89. — Round Timber Framed by Machine

resting-place on the shoulder of the post. As the tenon is only about two-thirds the depth of the cap or tie, when the caps and ties are in place, resting on the shoulder of the post, a mortise-like depression is formed on top of the post which is equal in depth to one-third the depth of the cap. In this mortise the shorter tenon, framed on the bottom of each post, is set. Fig. 87 represents the method of framing for "down pressure." To resist downward pressure the timbers are framed so that the posts butt directly together, the top of a lower post coming into immediate contact with the bottom of that next above, thus forming a continuous post from the sill to the top of the stope, or to a connection with the sills of the level above.

Not infrequently, as mining proceeds, it is observed that the pressure is being exerted from the side rather than from overhead. In such an event it is the practice to so frame the timbers that the caps form a continuous line from foot to hanging wall, as shown in Fig. 88. Fig. 89 shows the result of framing round timbers with a framing machine. In some mines round timbers are used exclusively, in square-setting as well as in drifts and elsewhere. Undoubtedly, round timbers are better than those that have been squared by saw. No round timber should be placed in a

Fig. 89a. — Stope in a Utah Mine showing Combination of Round and Square Timbers in Square-Sets

mine, however, without first peeling off the bark. This should be stripped at the time of cutting, in order that the timber may season properly. In some mines a combination of round and square timbers is used in square-sets, which has been found to work well. Notwithstanding the fact that the round timbers are never exactly the same size, the framing machine removes all inequalities and leaves the timbers cut uniform, so that they are interchangeable even with square timbers.

There are several so-called systems of framing timbers for square-setting — modifications of the two methods here illustrated for top and side pressure — and these differ from those shown only in the manner of cutting the posts, caps and ties. The primary idea of these is to reduce to a minimum the tendency of the timbers to split under pressure. The idea is all right in itself, but these elaborately framed sets resist pressure little if any better than those framed more simply, as shown. In the earlier years of the square-set, say thirty years ago, it was the practice to employ square timbers the dimensions of which were 18 × 18 up to 24 × 24 in., but even these massively timbered stopes caved. On the "Homestake Belt" in the Black Hills, the Father de Smet, the Deadwood, the Golden-Terra, the Caledonia, the Highland, and the Homestake mines each had large stopes filled with an elaborate square-setting of massive timbers, and yet great caves occurred in every one of these mines. Caves were of frequent occurrence on the Comstock, in the mines of Butte, Montana, and at almost every other place where square-sets without filling have been employed, both in America and abroad. In more recent years the tendency has been to reduce the size of timbers and to fill the stopes with waste as rapidly as possible. The timber sets, properly employed, make it possible to do this, and it is the only way the ore can be safely removed from large veins where the square-set system of timbering alone is depended upon. The increasing scarcity of timber has caused a corresponding increase in its cost, and mine managers have therefore sought to devise some method of mining which would admit the removal of large masses of ore in safety with the employment of a minimum of timber and at as low a cost as possible. This has led to the introduction of several ingenious schemes, designed to meet particular requirements.

Construction of the Square-Set in the Stope

The basis of the square-set system of timbering is the sills. These are placed on the floor of the stope and upon them is built the superstructure of posts, caps and ties. A system that is complete in its details includes not only the lower or mud-sills, but sills running at right angles to these, known as streak-sills or girders. It is upon these latter that the posts are placed. In many mines the girders are dispensed with and sprags, or tie-sills, are employed in their stead. Generally speaking, these latter fulfil every requirement. In such a case the posts set on the mud-sills. The chief reason for placing the sills is, not to form a base upon which to build the sets, but to provide a safe means of connecting one level with the next above. Where sills have been omitted, much difficulty has been encountered later in making connection between levels. As this is the main function of sills, it is apparent that sills should be as long as consistent with convenience in handling — at least the length of two sets, three is better, so that when breaking through from below, the sets of the stope above may be held securely until connection can be made between the timbers of the two stopes. This performance, of course, assumes that the work is taken up in small sections. Over the sills should be laid a floor of old timbers, lagging, and similar pieces of lumber to hold filling, or ore, as the case may be.

In starting a stope, when sufficient ground has been broken to afford the necessary room, the sills should be laid, their position having been properly determined by the mine surveyor, with reference to the sills in the stope above, which will be of advantage when the time arrives to connect the two levels. Where girders are dispensed with, the sills are laid at right angles to the strike of the vein — from foot to hanging — the same as the caps. At a certain distance from the wall, or at a point determined by the mine surveyor, a dap 1 in. deep is cut in the sill to receive the post, and at regular intervals other daps are cut, the distance apart being determined by the dimensions of the system, usually 5 ft. from center to center of the posts. The height of posts on the sill floor is often greater than that of those above, in order to give the necessary head room. In wet mines, too, provision is often made for drainage by giving the sill floor the necessary grade toward the shaft or toward some other drainage point, the sills conforming to this established grade, while the first set of caps above will be laid level. This makes the posts of the sill floor shorter as the stope proceeds away from the point of drainage. In most mines, however, no attention is paid to such refinements of practice, as the very slight tilt given the timber sets by the grade of the floor is negligible in considering stress.

When sufficient space has been made on the sill floor, and a double line of sills laid, the first four posts are placed in position and held there temporarily by nailing light braces, such as 2 × 6-in. scantling, or a strip of lagging, to them. The caps are then lifted to place and similarly secured temporarily, when the ties are put in. As soon as the first four posts, two caps and two ties are in position, they are spragged to the surrounding rock — sides and top — and made as tight as possible by wedging. If the shooting is heavy and the timbers small, they may be protected from the blasts by piling old timbers in front of them. Not infrequently a round of heavy shots will knock down two or three sets that have not been rendered sufficiently secure. The breast of the stope may be carried several sets wide along the vein — sometimes extending from wall to wall, if the vein is not too wide, but in very wide veins (those over 40 ft.) it is the better practice to carry the stope from foot to hanging. In fact, the block system is advised in all cases, and experience has shown that stopes may safely range from 40 to 60 ft. wide, depending on the character of the ground.

When the stope has become several sets wide and long, stoping may be carried upward. Sometimes it is necessary at first to build temporary platforms between the sets to bring the men within drilling reach of the ground, but as stoping proceeds upward, an effort is made to break the ground in such a manner that it will be unnecessary to build platforms. An experienced foreman will see that the machines are set and the holes drilled so that after blasting there will be a suitable place to set up for the next round. It is an easy matter for blundering miners to put in a round of holes which, after blasting, may leave the back too high to be conveniently reached without building cribs on which to stand while drilling the next round. Throughout stoping, when the men work on the timber sets, foresight is constantly necessary: always look ahead to see what will result from a contemplated round of holes, and decide what shall be done next. If this simple precaution be not taken, much of the time men will be working at a disadvantage.

In all places where large ore bodies are to be extracted and the stopes timbered with square-sets, it is advisable to cut raises from the sill floor of each stope upward to the floor of the level above, these raises to be lined with poles or plank and used as timber chutes or slides, down which all timbers to be used in the stope may be sent from the level above. Machine drills and other heavy articles may be sent down through these chutes also. These timber chutes may be cut much flatter than those cut for the passage of ore, or rock, as low as 30° being permissible, as the timber is thrown into the chute from a timber truck, and slides rapidly down to the floor where it is to be used. If the stope is carried up in the vicinity of the raise, so that the topmost floor connects with it, timber can be delivered by gravity to the level of any floor of the stope.

Chapter XXIII

MODIFICATIONS OF THE SQUARE–SET SYSTEM IN CALIFORNIA MINES

Although the square-set system of timbering is based upon well-recognized principles, these principles are not always faithfully followed. Some men appear to believe, judging by their work, that any kind of arrangement of timbers will be satisfactory and will answer every purpose. It is this erroneous idea that has led to the introduction of all sorts of modifications of the original idea of Philip Deidesheimer. Some of these modified systems are very strong, and do hold heavy ground as well, perhaps, as any other system; but after all they may possess no advantage over what may be called the Standard Deidesheimer system while they may include elements that render them more difficult of construction, requiring more men and more time to put in place. Among the California miners who usually did things "in their own fashion," no matter what others did, was the late Alvinza Hayward. That gentleman was the fortunate owner at one time or another of several of the largest and most valuable gold mines in California. Among them were the Union mine in El Dorado County; the Empire and Pacific (Plymouth Consolidated) and the Eureka and Badger (Amador Consolidated), in Amador County; the Utica-Stickle mine in Calaveras County; and a number of others; but each of those mentioned were large and rich mines, and in them Mr. Hayward introduced his own methods. These methods, while differing in important particulars from those in common use elsewhere, were nevertheless satisfactory, and apparently served every purpose. However, we believe that the Hayward systems were more cumbersome than the Standard system, and otherwise objectionable in some of their features.

One of the main departures from the usual practice was the placing of the posts directly upon the rock floor of a level, such a thing as a sill never being employed. The posts were 16 ft. in height, and while the framing was essentially similar to that employed in the Standard system, it necessitated the wedging in of sprags about the middle of the posts, 8 ft. above the floor. The sets, when in position, were very strong, being made up of posts 24 to 30 in. in diameter, and caps 20 to 24 in., with ties 12 to 24 in. diameter. Such huge timbers should support ground if any timber would, but even these stopes caved occasionally where too large an area was stoped. In those early days filling was not extensively practised. The main disadvantages, however, seem to have been the great size of the posts and the corresponding difficulty in handling these great logs and placing them in position. Moreover, the system required a stope to be nearly 18 ft. high before these great posts could be set up, and in many places this was accomplished with danger, owing to the size of the unsupported excavation that had to be made before the timber could afford any support to it whatever. Of course, the difficulty of connecting levels was not simplified by the omission of sills. In time the practice of using 16-ft. posts was discontinued for 8-ft. posts, which at once made the Hayward system in most respects similar to the Deidesheimer system.

In the Utica mine, at Angels, another modification of the square-set system was introduced, and this was followed to some extent by other mines on the Mother Lode. Strange as it may seem, not a mine on the Lode, with its great bodies of ore, where square-setting was commonly employed, made use of sills prior to 1893. At that time, or in the following year, sills were first introduced in the Utica mine, but few of the other mines followed the example of the Utica. It was claimed that the sills would be of no use, as they would rot and become useless before the levels were connected. This was, in most cases, true, owing, as previously explained, to the method of working these mines. The veins or ore bodies were often of great size — 100 ft. or more in width — while the milling plant was generally small, consisting usually of 40 to 120 stamps. A body of quartz 100 ft. high, 100 ft. wide, and 100 ft. long contains nearly 80,000 tons. A 40-stamp mill crushing 3000 to 4000 tons a month could run steadily on this mass of ore for two years or more, and as stoping was generally progressing on several levels at one time, it required several years to carry up one of these great stopes of large section from one level to the next above, during which time the sills had ample time to rot and become useless.

The old methods of mining in most of the great Mother Lode mines are antiquated, and although cheap for the time being, were not the best when the economy of the future is considered. Each manager sought to stope his best ore and to make a large output. The result was large excavations, generally without filling, and ultimately caves, with the attendant expense and danger of recovering such ore as could be drawn from the caved

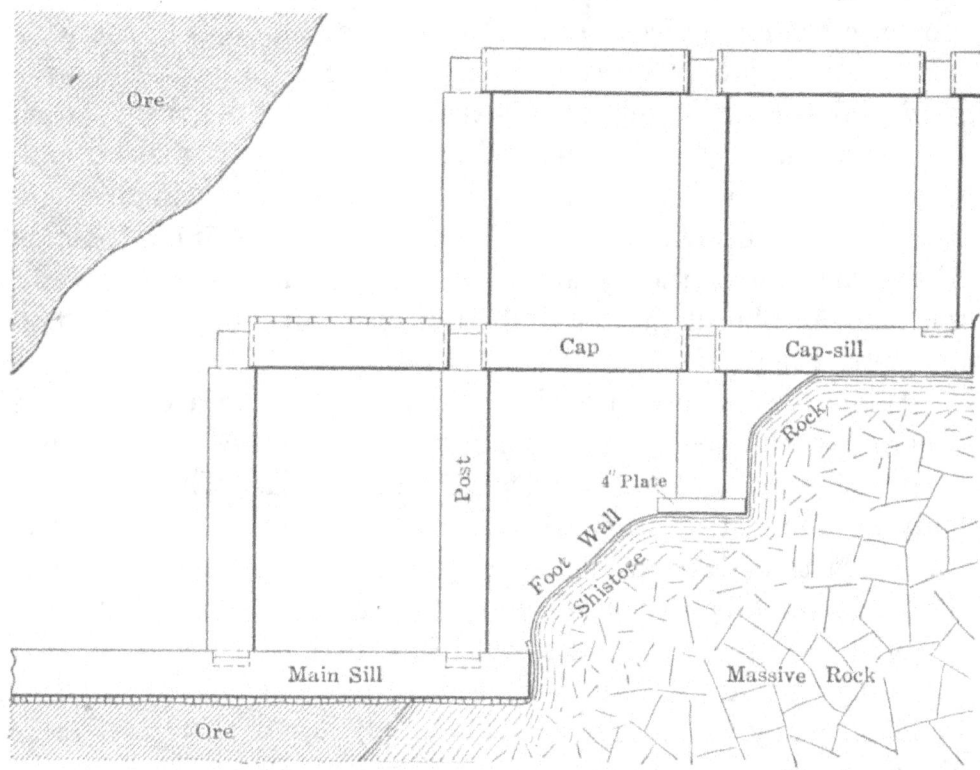

FIG. 90. — Use of the Cap-Sill in Square-Sets

ground. In some mines the posts have been set up from 8 to 16 ft. in height and the caps placed on top of them, completely covering the top of the post, while no framing was done to receive the foot of the post of the next set above. This necessitated much work with ax and adz in the stope, something which should be avoided as far as possible. In other mines the ties or sprags have been omitted, probably with a view to economizing timber and time, but this does not look like real economy, for if the ground shifts slightly, the timber sets are very likely to ride sideways and "jackknife," as the miners call it, causing complete collapse. When heavy standard square-sets do this under unequal pressure, it is scarcely to be expected that a system of timbering from which important members are missing will do any better under similar circumstances.

The great variation in the angle of dip of the walls of ore deposits and veins, as well as in the character of the walls or surrounding rocks, often necessitates the introduction of special methods of timbering, such as are illustrated in the accompanying sketch, Fig. 90. In this case the foot-wall was too soft (talc schist) to afford firm support to the foot-wall posts, and in consequence the miners were obliged to cut deep hitches in the foot-wall rock in order to secure a firm footing for posts, but the particular feature of this case is the introduction of what is called the "cap-sill." It is at once apparent that this is most useful, wherever it may be necessary to meet such conditions as are described. The name "cap-sill" indicates the peculiar function of this member of the square-set system of timbering. Cap-sills may be introduced in the timbering of small stopes as well as in those of large size.

Among the large mines of the West, where round timbers are exclusively employed in supporting stopes, is the Utica-Stickle mine, at Angels, Calaveras County, California. The main ore body worked in this property ranged from 40 to 120 ft. in width, was more than 1000 ft. long, and about 400 ft. high. The formation is amphibolite schist, which on the foot-wall side is altered to soft talc, often wet, and several feet in thickness. Generally, the hanging-wall is solid, but often came away in great slabs weighing many tons. Obviously, the stoping of this great mass of ore, which stood at an angle of nearly 80° on the hanging side, but with greatly varying dip on the foot, required much skill and ingenuity on the part of the mine management, as it involved no little danger. The accompanying sketches, Figs. 91 and 92, will give an idea of the method of timbering employed in the Utica-Stickle mine. The timber used is round, peeled pine logs, delivered under contract at the mine. I have been informed that the cost of these logs was 10 cents for each inch in diameter, for a 16-ft. log. Thus a 20-in. log would cost $2, while one 24 in. in diameter would cost $2.40. The timbers are mostly over 18 in. diameter and are as large as 24 and even 30 in. Sprags are mostly 12 to 16 in. diameter. The larger timbers are all

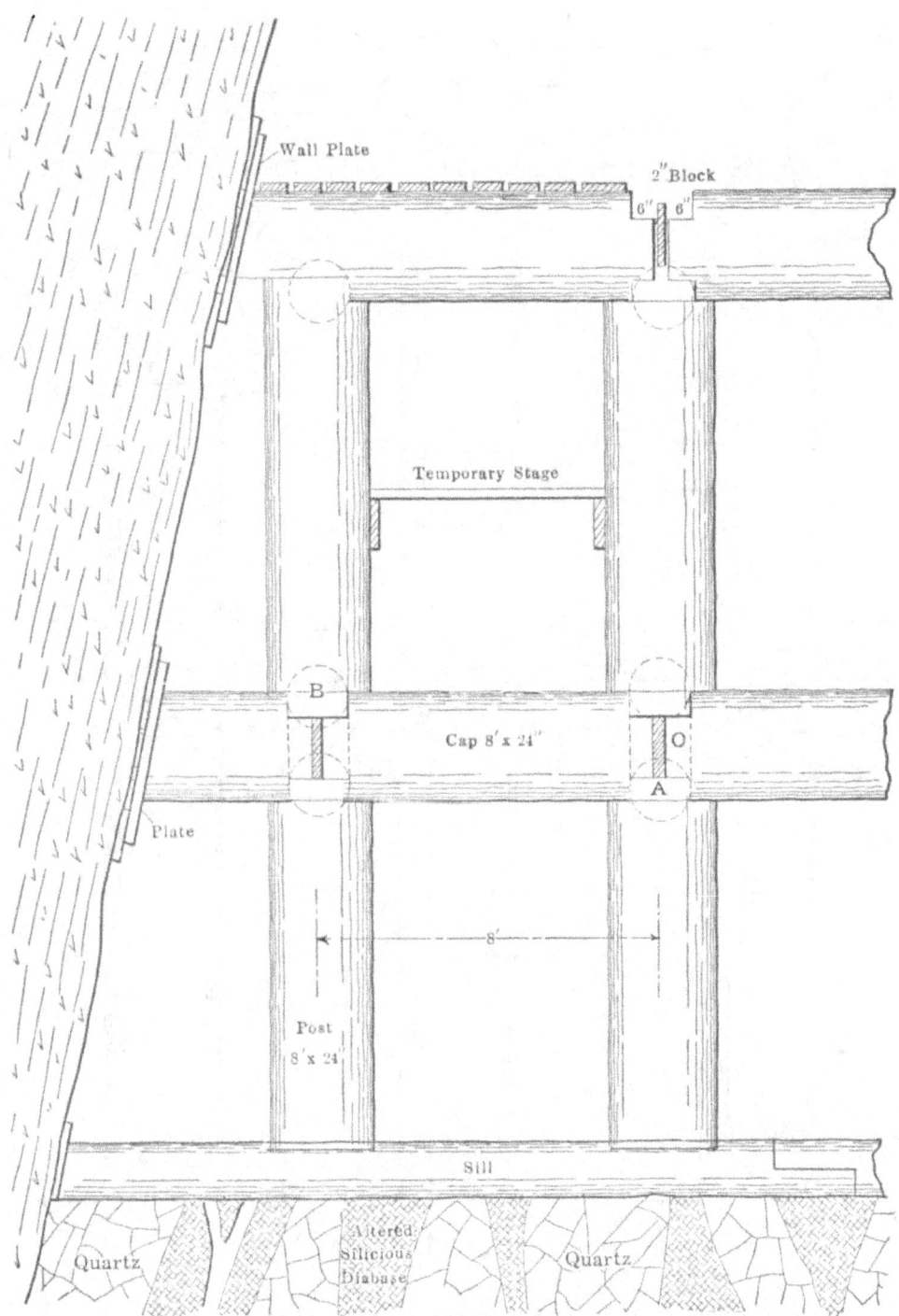

Fig. 91. — Utica Method of Square-Setting. (Cross-Section)

sawed into 8-ft. lengths and framed by machine, the tenons being 14 inches square, top and bottom, and 4 in. long. Caps are

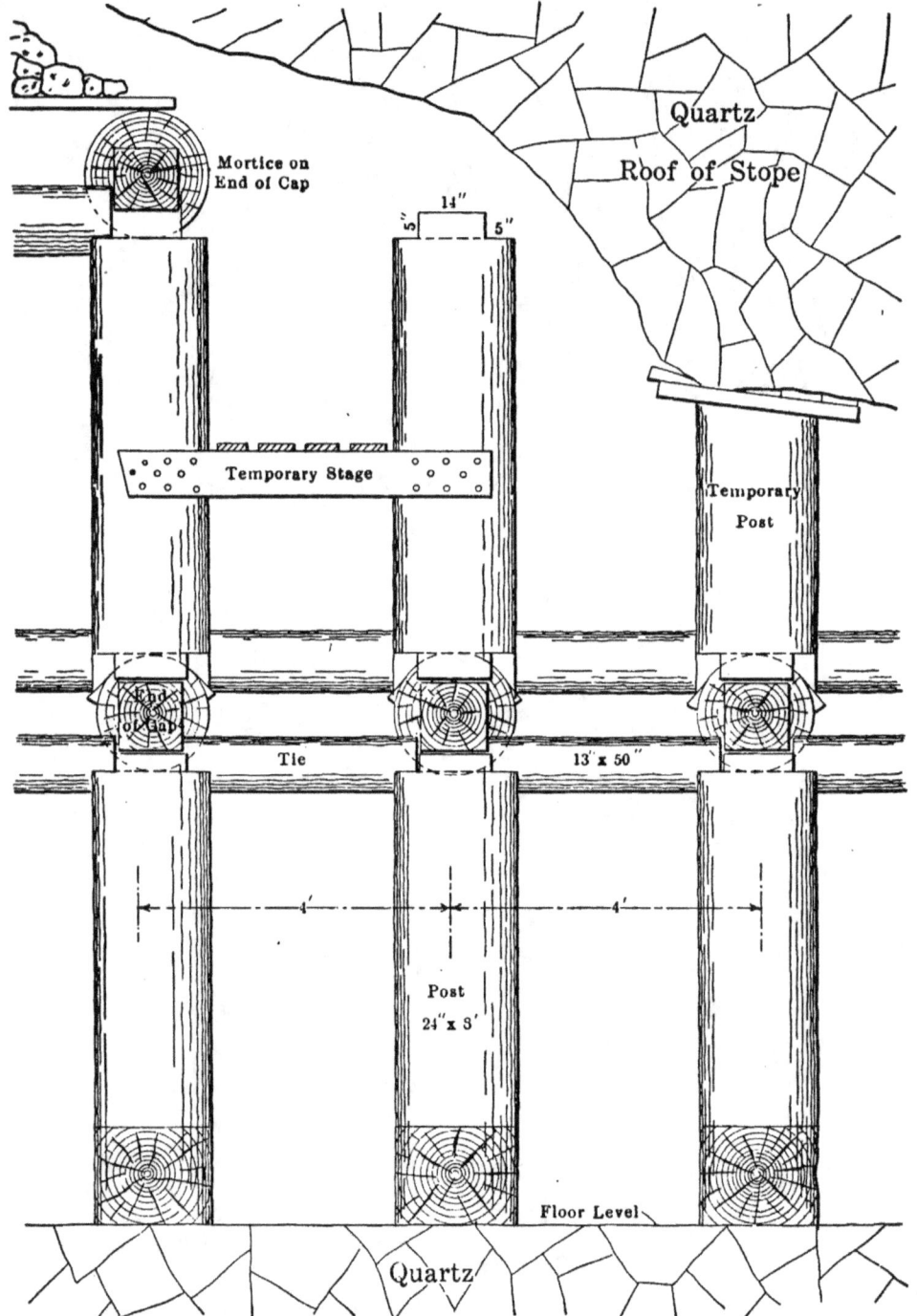

Fig. 92. — Utica Method of Square-Setting. (Longitudinal Section)

framed to 14 in. square, the "horn" being 6 in. long. It will be observed that when the caps are placed on the posts, the tenons or horns fail to meet by 2 in. This space is filled with a piece of

2-in. plank 14 in. square. Some years ago the men became rather careless in setting up these timbers, and in at least one stope the 2-in. plank was omitted, with the result that when pressure was exerted the caps gradually butted together and the hanging-wall settled, many heavy slabs falling and causing numerous accidents as well as much additional expense to hold the ground, which should never have been allowed to start.

At the junction of caps and posts it will be observed that two sprags (ties) are introduced to hold the upper and lower post in position independently. The sprags are 4 ft. long, the lower one being framed with a horn 4 in. in length, as shown, on its upper half. These projections rest upon the shoulders of the post, while the upper sprag is tightly wedged between the posts. In this mine the caps and posts often weigh 700 to over 1000 lb., and the timber gang is made up of powerful men who are able to handle these great logs with comparative ease. The timbers, when in place, form a very heavy and strong system of support, but seems to possess no advantages over the standard sets. Owing to the impossibility of procuring round timbers that are uniform in diameter, this system involves a great amount of work in the mine, dressing timbers with ax and adz making a great quantity of chips, which later prove a nuisance in the mill.

At the Wildman mine at Sutter Creek, Amador County, California, the method of timbering is similar to that at the Utica. The posts are provided with tenons 6×12 in., which are $8\frac{1}{2}$ in. long on the upper ends and $3\frac{1}{2}$ in. long on the lower ends. The posts abut upon each other, the caps being framed with a tenon 12×12 in. and 4 in. long. These tenons rest upon the shoulders of the posts. Double sprags are also in use at the Wildman, but are placed side by side, so as to catch the two caps, instead of one above the other, as in the Utica system. In the Wildman mine the timbers are generally smaller than those used at the Utica, being from 14 to 18 in. diameter.

Construction of Chutes in Square-Sets

In most mines employing the square-set system of timbering ore chutes are built in the sets of the sill floor for the purpose of drawing off the ore broken in the floor above. When the excavation has proceeded sufficiently to admit of the timbers being carried two sets high, tracks may be laid wherever desired on the sill floor, and ore bins built in the timber sets, with chutes arranged along the car tracks at convenient intervals in order that cars run beneath the chutes may be expeditiously filled. It is the usual practice to lay a heavy bulkhead on the upper floor of the sets, so that when blasting the rock may fall on the bulkhead, where it can be broken up with hammers to a size that will readily pass the chute doors. If the ore is not drawn off from the bins as fast as broken, the bins will soon be filled to the top, when the ore may be blasted down directly upon the broken ore, where the large pieces may be reduced to the desired size. The ore protects the timbers of the set, and the bulkheads become no longer necessary. Where large stopes are completely filled with square-sets, however, this scheme of carrying a large amonut of ore in the bins is not the best practice, as it exerts too much weight on the timbers adjacent to the bins, while there is no corresponding weight elsewhere, which sometimes displaces timbers and throws the sets out of line.

The difficulties encountered in the endeavor to lessen the cost of extracting large ore bodies have been numerous, and many novel schemes have resulted. There has been a growing tendency to reduce the amount of timber required in mining large ore bodies, and in many instances these efforts have met with success. In others the amount of timber used is so small as to be almost negligible.

Chapter XXIV

STOPING LARGE ORE BODIES BY THE BLOCK SYSTEM AT BROKEN HILL, NEW SOUTH WALES

One of the great lodes of the world is that at Broken Hill, New South Wales. It is a huge deposit of lead-silver ore, with much iron, zinc and copper sulphide. The lode is situated in the Barrier Range of mountains, and at the time of its discovery was many miles from the nearest settlement and about ninety miles from the nearest railway, in the midst of a desert not unlike that of Southern California. The richness of the ores quickly brought Broken Hill into the foremost rank of great mines, and millions of dollars in precious and base metals have been taken from the great lode. As in other great lodes, the early mining operations were conducted with a view to realizing promptly the greatest possible profit and with little regard for the future. The result is that mining at Broken Hill is to-day dangerous, and owing to the high cost of timber and labor new methods were devised several years ago to work the caving, shifting ground with as great assurance of safety as it was possible to secure, and at the same time extract the ore.

Edwin K. Beaumont has described the mining methods at Broken Hill in the "Transactions of the Australian Institute of Mining Engineers," from which the following is abstracted: "The impressions of a new arrival on coming to the Barrier are anything but pleasant or reassuring, for when, after spending a whole night in rattling over 300 miles of almost desert country, one seeing the line of lode, about one and one-half miles in length, with its long chain of chimney stacks, poppet heads, engine houses, concentrating mills, and immense mullock and tailings dumps, etc., finds it hard to imagine that this is the famed Barrier Range or Broken Hill, unless one is a mining man who has been on similar fields. The original Broken Hill is now a thing of the past, having been entirely removed by the large open cuts, from which the oxidized ore is being extracted down to from 200 to

250 ft. The story of the finding of silver by the stockman on Mount Gipps station and of Rasp's shaft (Rasp himself being a station hand at the time), and of the mining of earlier years, are familiar, when only the oxidized ores of the upper levels of the lode were worked down to 300 or 400 ft., and, as on new fields, the methods adopted were crude, until with great advance of output and rush of population, more modern and advanced systems were adopted, upon the arrival of American mining managers and engineers with the square-set system of timbering, as carried out in ore mines of America and elsewhere.

"In mining these large ore bodies the square-set system is employed. This method of mining was introduced on the Lode by W. H. Patton, an American mining engineer, and an old Comstocker. The timbers are framed in the usual manner by machines in the timber shop. Some of the mines have their own sawmills, and buy their timbers in long lengths from the Port, and cut all timbers to templates. They are sent underground ready for the miners and timberman to frame up in the stopes. When the ore is hard and compact (and it is hard sometimes, especially in the class of ore containing rhodonite), I have frequently seen five drills blunted to bore a hole less than 1 in.; but when the ore is friable, it is then timbered close up to the working face (as shown in Fig. 93) on the upper floors. In this system the miners are always working close to the face or back, which they can easily examine to make sure of its safety. One disadvantage of keeping the timbers so close to the face is that frequently a heavy shot will 'throw,' and thus knock down several sets and shake others, thereby causing delay and rendering the working face unsafe until the timbers are re-erected, and there is always difficulty in securing them all as firmly as they were originally. Lagging of 10 × 2-in. Oregon pine are laid on each floor as the stopes rise upward to the next level, and chutes for conveying the broken ore to the sills and thence to the trucks are provided at convenient intervals, and slides placed to run the ore to the chutes from the working face, as shown on the plan and section showing the stope (Fig. 93). It will also be noticed that the end sets of each floor are wedged firmly to the foot and hanging walls of the ore body, and frequently notches or hitches cut to secure a solid bed. In theory, as a system by itself, these seem admirable, but in practice (without being filled with waste

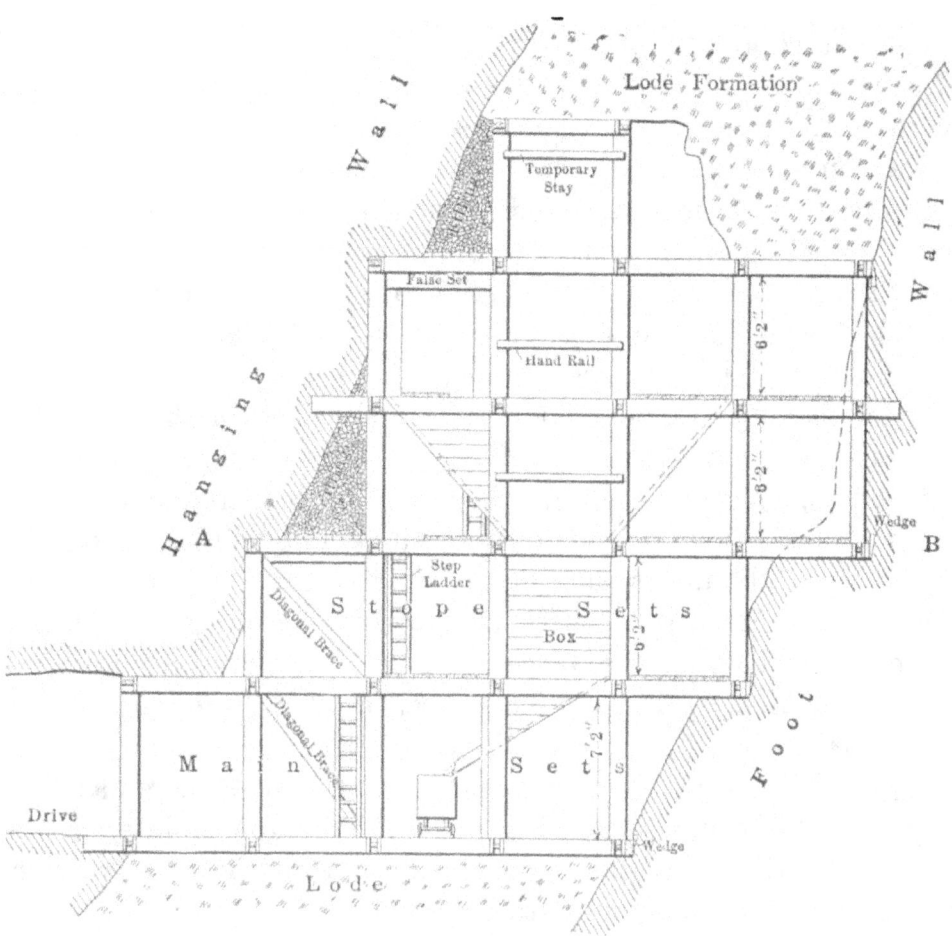
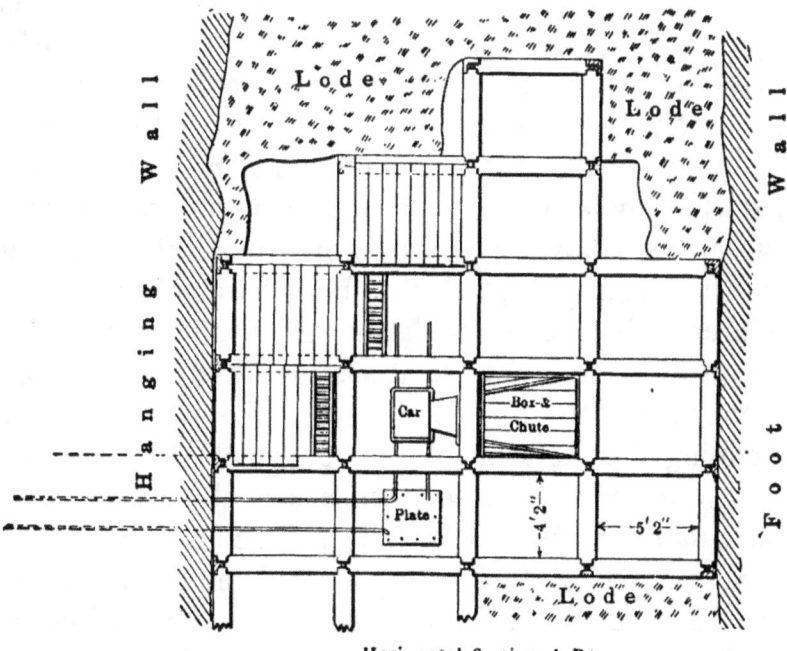

Fig. 93

as they now are) they fail, for after the ore has been extracted, any movement or pressure of the walls of the lode causes a collapse of the set. They have gained the weird name of 'creeps,' and a more complete state of chaos can hardly be imagined than a creep, that is, broken and splintered timbers, and masses of ore and mullock in one almost unapproachable mass, often rendering the further working of that portion of the lode impossible, and thereby losing large quantities of ore in the débris; but it is a noticeable fact that in those mines where the managers did not rely on the timbers alone, but judiciously filled in the sets with mullock from wall to wall (leaving only the necessary openings for chutes, gangways, etc.), when any movement came in the walls of the lode the timbers and surrounding filling stood the burden, and the mines were singularly free from 'creeps.' Some idea of the pressure on these timbers may be imagined when I state that I have seen a piece of 10 × 10 in. Oregon pine compressed to barely 3 in.; have also seen a 10 × 10-in. vertical leg driven $4\frac{1}{2}$ in. into the horizontal cap and sill at its ends, without bending the leg, and have frequently noticed, when there has been any lateral pressure, the 10-in. piece of Oregon splintered like a piece of willow on the convex side, and on the concave side, though bent one foot, was still unbroken.

"The life of timbers underground depends a great deal on location. In some mines I have noticed Oregon pine that had been in approximately ten years almost sound, while in others the same class of timbers, if put in a badly ventilated stope, in about three or four years had completely decayed by a kind of moldy dry-rot. In the upper portions of one mine I noticed a lot of joggled logs of blue gum, from 6 in. to 9 in. diameter (brought from the river in early days), and they were worm-eaten and quite rotten, while the mulga and black oak, both hard native timbers, were quite sound; but as these latter are usually only about 4 in. to 8 in. in diameter, they are almost useless for underground timbering, except as lathes, or for latticing the sides of square-sets and enclosing the mullock fillings, for which they are sometimes used. I omitted to mention, when describing the mullock filling of the square-sets, that 10 × 2-in. planks are used by some mines, but other mines, having their own sawmills, rip the 10 × 2-in. lath in halves and use the 5 × 2-in. lath, thus effecting a small saving in the amount of timber used; but I have noticed also that these lighter laths frequently give way when any great pressure from the mullock is thrown on them. Other great detractions to the square-set system are the great cost of timber and the liability to fires.

"When considering the amount of timber required to timber a lode, the total feet is enormous, and at, approximately, $38 per 1000 ft., the cost greatly reduces the profits. We have in the Central a width of over 270 ft. from foot-wall to hanging-wall at the 600-ft. level, the greatest width on the Barrier, taken out in blocks about 50 ft. wide right across from one level to another, viz., 100 ft. vertically. There is the liability to fires, of which the fire in Block 11, of about seven years ago, and the fire in Block 12, of four years ago, both of which are still burning, are examples. These cause great expense in extinguishing, and greatly hamper the working of the upper portions of the lode. So much timber of inflammable nature is a great menace, especially when the stopes are well ventilated by winzes from the upper workings. These winzes serve as vents or chimneys, and spread the gases throughout the mine, and the result is loss of life, as it was on both occasions referred to, when men went below to locate and attempt to overcome the fire. It will be seen that the deduction to be drawn is that this system is an admirable one when combined with filling; but it is also an expensive one.

"Owing to the sulphide ore requiring more costly and tedious treatment and preparation before smelting than oxidized ores, and the many and varied expenses incidental to the mining, due to the hardness and the work of extraction and handling, it was necessary that managers and others interested should devise some safer and cheaper methods, which would also reduce expense. Later methods have gradually evolved into their present forms; and their almost universal adoption on several mines on the line of lode — with sundry modification to suit individual cases — proves their efficiency. The square-set system is far from annihilated, as it is still used where applicable, especially on the sill floors where gangways are required; also chutes, outlets, etc. Here the sets are on solid bottoms; being well wedged against the hanging and foot-walls or the sides of the stopes, they are firm and permanent, and with the 10 × 3-in. planking to carry the mullock filling, they form convenient passages about the workings. So it is still used in conjunction with these later systems, as a valuable and necessary adjunct."

Underground Open-Cut and Sloping-Stope Systems, at Broken Hill

"The term underground open-cut system may at first seem erroneous or misleading, as the term open cut is generally applied to excavations from the surface downward; but the above name is that generally given by the miners to the large stopes which are worked under this system. The drives are first run along the foot and hanging walls, and then through the ore body. From the upper levels winzes are sunk and cross-cuts driven at convenient intervals. The winzes serve several important purposes. They ensure a complete and lasting ventilation to the stopes during their upward way by carrying off all noxious gases as they form in the lower workings or are given off by the sulphide ore, thus enabling the miners to work with a greater degree of comfort and also to do a fair shift's work, which could hardly be expected in a hot stope with a constant atmosphere of about 90°. The winzes serve as passes or chutes through which the filling is conveyed from the upper levels, and, by a succession of chutes and winzes from the surface, deposited where desired. This system is successful in the Central mine, for the mullock or waste is broken in a large open cut on the surface and is conveyed in side-tipping trucks, of a capacity of one cubic yard, drawn by horses through a tunnel, then discharged into a chute, from which, by a series of winzes, chutes etc., it is distributed throughout the mine where required. The winze is used as a starting place or face from which to work the stope, and, after the ore is extracted, say, the first 10 or 20 ft., it is timbered up closely into two compartments, as shown in Fig. 94. One compartment serves as a chute or pass for the ore to the sill floor, as the stope works upwards, and the other compartment as a ladder way and means of ingress and exit for the miners and others to the upper workings of the stope. The sides of the initial drives on the sill floor are extended to the desired width along the lode, and thus the stope is formed on the sill floor, the sill timbers placed in position and then filled up. On top of these timbers the bedding for the filling is placed, as shown in the section, Fig. 95, being 10 × 10 in. and 10 × 3 in. timbers, arranged to carry the burden. Above these 'sollars,' as they are called, the only timbering is that of the chute and ladder way, all other spaces being filled in with mullock from wall to wall, as indicated, which is placed in layers of 7 to 12 ft. As the broken ore falls, and the traffic also is all on the mullock filling, each succeeding layer gets well rammed and solidifies before the next one is placed on it. In the large open stopes in the Central mine, almost all the boring is done by machine drills driven by compressed air. These

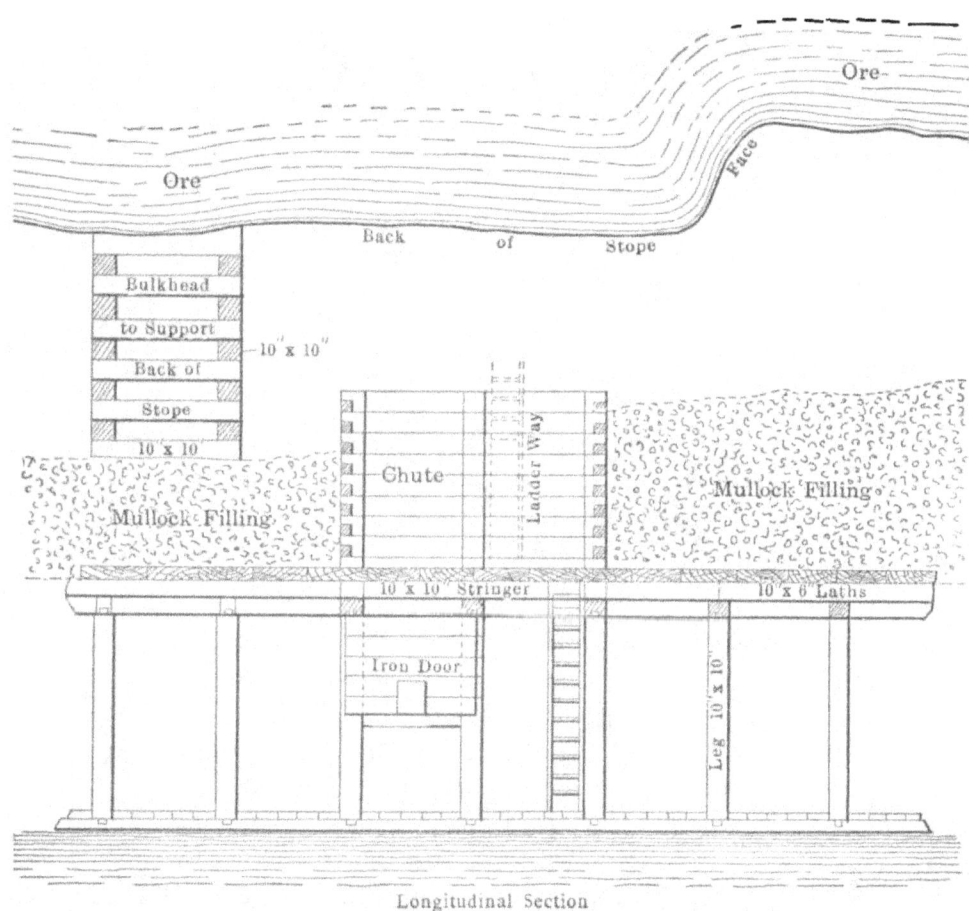

Fig. 94. — The Underground Open-Cut System

bring down the ore in large pieces, frequently from 7 to 8 ft. by about 2 ft. wide. These pieces are then bored by hammer and drill, and popped into smaller sizes, then sprawled into sizes — generally less than 1 ft. long — for throwing down the chutes and removal in the trucks, which are all end-tipping and hold about 1600 lb. of broken ore. When the 'back' or top portion of the stope is 'heavy,' or seems dangerous and likely to come away, bulkheads are built under it. These consist of 10 × 10-in.

timbers placed at right angles to each other, one above the other and tightly wedged. (See Fig. 94.) When bulkheads are built on the mullock filling, a bed of 10 × 4-in. sollars is first laid on the mullock, to distribute the pressure over as large area as possible; then the first 10 × 10-in. timbers forming the bulkhead are laid

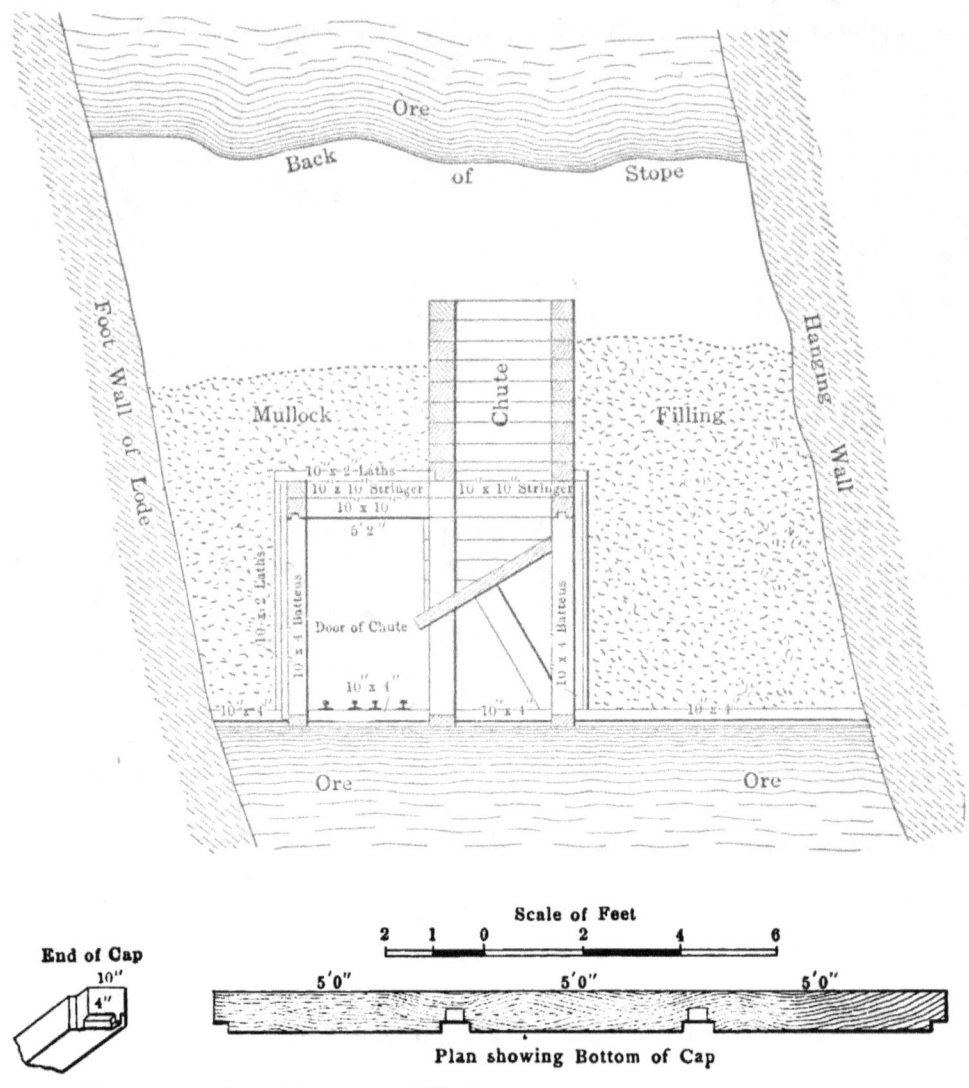

FIG. 95. — Cross-Section of Underground Open-Cut System

transversely across the sollars. These timbers are afterwards removed, the burden shot down, and the same timbers used over and over again.

"A somewhat similar modification of the same system, called the 'sloping-stope system,' is shown in Figs. 97, 98, and 99. This method is extensively used on the Broken Hill Proprietary

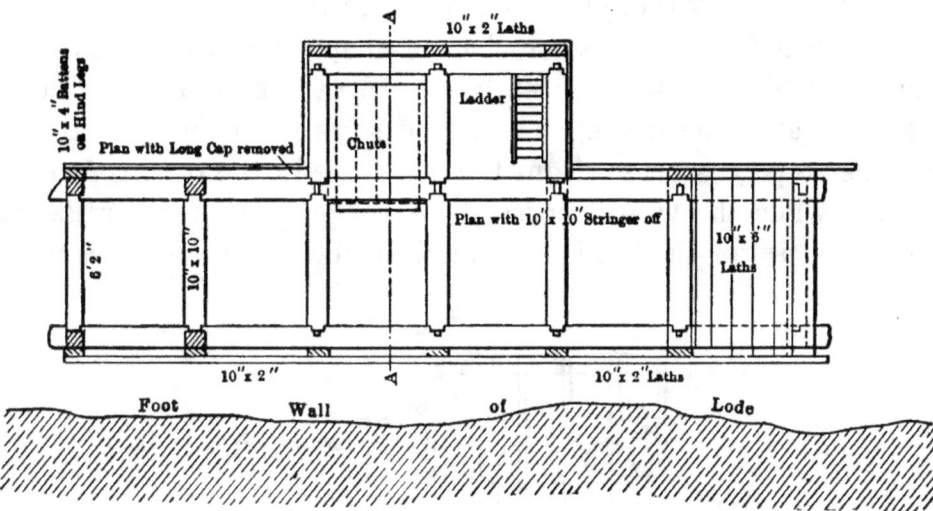

Fig. 96

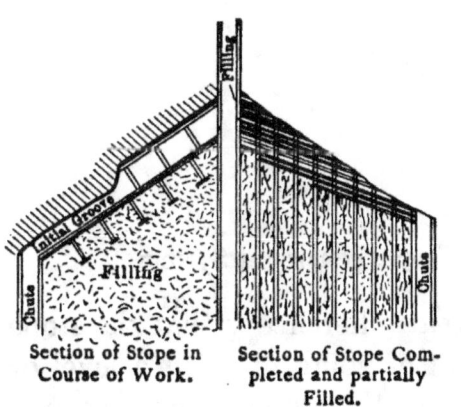

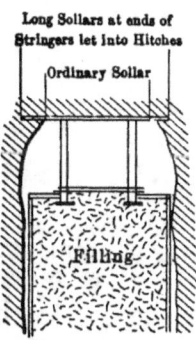

Stoping in Horizontal Layers.
Adopted in Hard Ground.

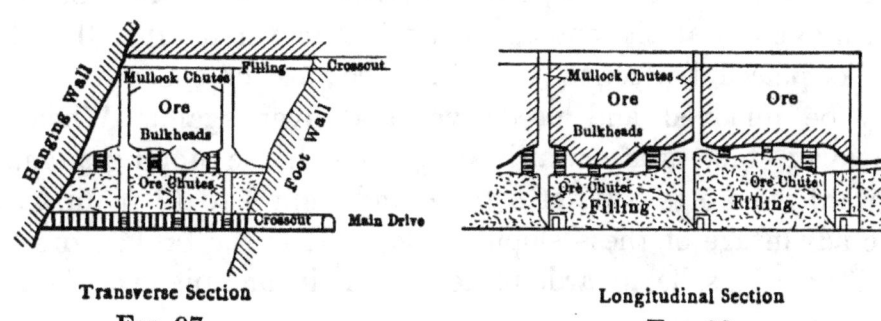

Fig. 97   Fig. 98

mine, and I am indebted to E. J. Harwood, C.E., mining manager, for permission to copy his drawings showing this system. In many instances the same description will apply as in the foregoing notes on underground open-cut stoping, viz., the levels on the sill floors are first formed, taking notice that the width of the stope depends on the nature of the ore to be mined, or its ability to support itself by leaving the back in the form of an arch; the whole stope, when formed, is somewhat in shape like an isosceles triangle, of which the level or sill floor forms the base and the winze the apex; also, the winze, as before, is sunk from the level above, and the stope is started, from the winze, as in the other open-cut system, while the winze serves the same purpose, for ventilating and as a pass for the mullock filling into the stope, also as a chute

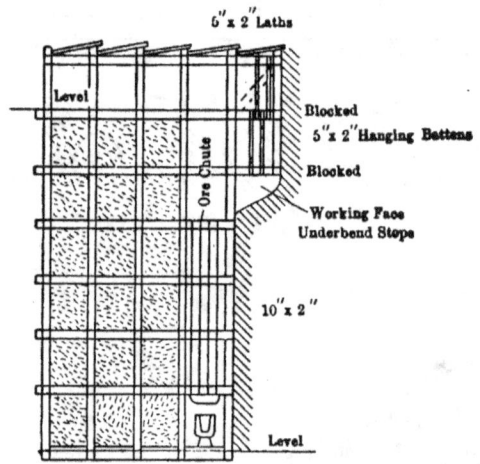

Fig. 99. — Vertical Section of Stope, Square-set System

for the ore to the sill floor. The great difference is that the stope slopes laterally to each side, instead of going up with a level or even floor, and as these sides rise with the stope, provision must be made for preventing the mullock filling from running into the adjoining stope when it rises. This is done by placing vertically at the sides of the stope, about 5 ft. apart, 10 × 4-in. stringers which overlap at the ends, and are then covered with 10 × 2-in. planks placed horizontally against the face of the ore. These may be removed and used over and over again. When the adjoining stope is afterward being worked, the stoping advances forward from 5 to 8 ft. at a time, and from 8 to 12 ft. upwards. The advantage of these sloping sides is that the broken ore falls on 10 × 4 in. sollar boards placed on the incline plane of the mullock, and thus rolls to the chutes at the sides without further handling, excepting, of course, the large pieces, which require hand boring, popping and spawling, as before mentioned. It will be noticed on referring to Fig. 97 that the stope is filled in with mullock to within 2 or 3 ft. of the back, and the stope is always worked downward, starting from the winze; 10 × 10 in. legs rest on the sollars (or on bed logs left in the mullock), or at right angles to the sollars and also to the back, as they are nearly parallel; these are tightly wedged and blocked and only removed as the ground is taken out. When the stope is about 9 ft. high, the sollars are taken up and stored aside for further use. The stope is mullocked up again to within 2 or 3 ft. of the back, when the chutes are again built up a porportional height and the sollars replaced on the mullock, the sloping process, as before, taking another slice from the back, also starting from the winze downward. The Broken Hill Proprietary Company has of late years been adopting a modification of the square-set system in working these 'sloping ores' by timbering up the middle of the stopes with square-sets, which are filled with mullock as the work proceeds; but as the back — or working face in this instance — is sloping, as in the last mentioned method, each successive floor of sets stands back one, or in such a manner that the outside faces of the sets follow as nearly as possible the same angle of inclination or inclined plane as the face of the ore body. In this way the miners are always within a safe distance in working and examining the face and back of the workings, and all the favorable points of the other adaptation of the sloping-stope system apply to this system, with the advantage that the miners have a good footing on the set timbers, and the great convenience offered for the despatch of the ore through the chutes constructed in the square-sets."

Another modification of the square-set system as adopted in the Central mine is known as the "block system" (see Figs. 100 and 101). The lode for its entire length through M. L. No. 9 has been surveyed into parallel blocks each 50 ft. wide — *i.e.*, ten 5 ft. wide sets. Each alternate division is a block and the next a stope. The whole level is gradually developed by a drive along the foot-wall and by cross-cuts to the hanging-wall, thereby determining the width of the lode along its entire length, and the stopes are then carried from the foot-wall to the hanging-wall on the sill floor and the space filled with square-sets, leaving every facility for forming the necessary gangways, chutes, etc.

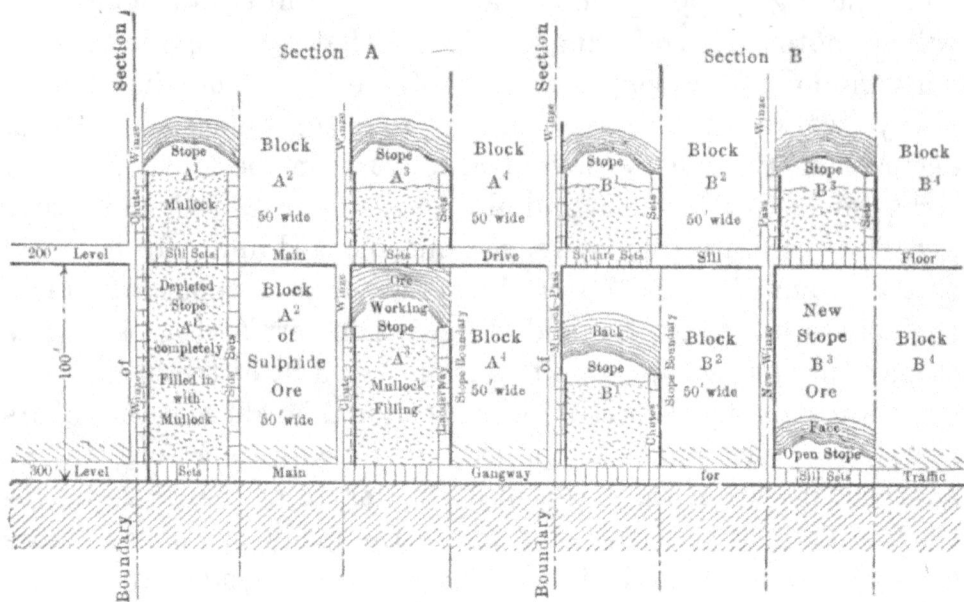

Fig. 100. — The Block System of Stoping. (Vertical Section)

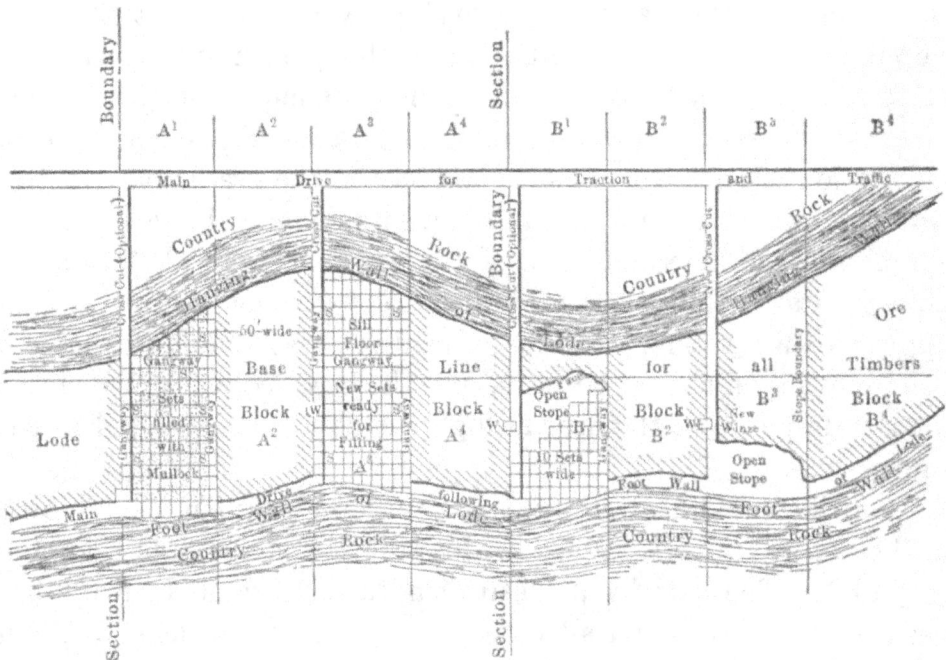

Fig. 101. — The Block System of Stoping (Horizontal Section)

These are then filled in with mullock, and the stope starts on its course upward, being exactly 50 ft. wide, the entire width of the lode at that point, thus leaving a pillar of ore 50 ft. wide in each side of it from wall to wall, which will carry all pressure during the mining of this stope. A run of square-sets is put in each side of the stope as it goes upward, forming a gangway and ladder way, the sides of which are lathed or paddocked off, thereby confining the mullock filling in the center of the stope. The ore is broken by machine drills driven by compressed air, and in the same lifts and proportions as in the before-mentioned open-stope system. The ore falling on the mullock filling in the center of the stope is popped and spawled into suitable size for handling and trucking to the shaft for haulage to the surface. One great difference in this system is that the winzes — 6 × 5 ft. — are always sunk 100 ft. apart at the side of each alternate stope, being half in the stope and half in the adjoining block, thereby saving a sceond winze when the block is being taken out at any future time. The ore from the adjoining stopes having been all extracted, and the space filled with the mullock, this winze will then be available and serve the same purpose for the remaining block. The same advantages of ventilation, mullocking and stoping all apply to these stopes, as in the foregoing open-cut and sloping-stope systems, and they are mullocked up in the same manner, excepting that the chutes for conveying the ore from the working faces to the sill floor are located in the runs of sets, placed on the side of the stopes for that purpose, and the chutes can be constructed at suitable intervals for the workings.

The advantages and disadvantages of these several systems are: The whole of the ore body is, or at least can be eventually, extracted, and after extraction of lode material comparatively few if any large voids or openings are left; which also leaves the surface areas for works, mill, machinery, etc., almost free from risk of subsidence. The great advantage evident from the presence of mullock filling — in lieu of a forest of timber — is the immunity from risk of fire. The miner is always in reach of the "back," *i.e.*, from 3 ft. to 9 ft., and can readily sound and examine the back of the workings and make sure they are safe; this secures a greater freedom from accident through masses of ore falling on men while at work immediately under them. Unfortunately, however, recent experiences have shown that though a place may be sounded and examined by miners with a lifetime experience and reported as safe, yet a few hours afterward the back may fall in and reveal a fault or crack which the sounding did not make known, serious or fatal accidents resulting. But even then this cannot in any way be compared to the great risk incurred by men when re-erecting square-sets that have been knocked down by a heavy shot, for sometimes a charge will bring away more ground than anticipated, and a dozen or more sets will come down, and the men will have to work under a probably dangerous back in re-erecting the sets or staging before they can actually examine it and assure themselves of its safety.

Another advantage is the saving in the expense of timber. Of course, against this must be placed the cost of quarrying the mullock filling on the surface and conveying it to the stopes, which would, however, be required in any case in filling the square-sets in the other systems. The great advantage of good air is important, as the mullock fills in all spaces except the winzes and that part of the stope that is being worked, thus ensuring always at the working face a current of air, which both carries off the smoke after firing and adds to the miner's health and comfort, removing much that in former days made the miner's life a hazardous and unhealthy one.

There is one other important method employed, by which a large amount of oxidized ore has been removed. I refer to the open-cut system. The large surface excavations are one of the chief sights of Broken Hill, and though descriptions may give a slight idea of their extent, I think they must be seen to be understood or appreciated. Imagine an open cavern, three-quarters of a mile long, traversing the whole of the Broken Hill Proprietary Company's blocks. The widths of these cuttings vary from 120 ft. to about 350 ft. There is also a width of 300 ft. opposite McGregor's shaft, in the center of Block 11. The cuts are down about 250 ft., and they are recovering a large amount of timber that was used underground in the square-set system of stoping in the old 200-level workings, the ore from which was then hauled up the various shafts before the open cuts reached their present depths.

A method by which the greater part of the ore is raised from the open cuts to the surface is by means of what is here called the "Flying Fox" — a large skip which is hauled up along an aerial ropeway and thence discharged into large ore bins at the sides of the railroad lines on the surface, from which it is conveyed to the mill or to the smelters. A mast is erected on each side of the cut and a cable stretched over an iron saddle near the top. The cable is anchored securely at either end, while on the surface is located the hauling engines, having a loose pulley and reversing gear. An attachment called a bicycle runs along the main cable across the cut, having on it four pulleys. The upper two travel along the cable, the lower two being used in hoisting the skip vertically from the cut; the rope, called the traveling rope, then draws the bicycle, and, of course, the skip with it, along the cable. When it is over the bin on the surface a self-

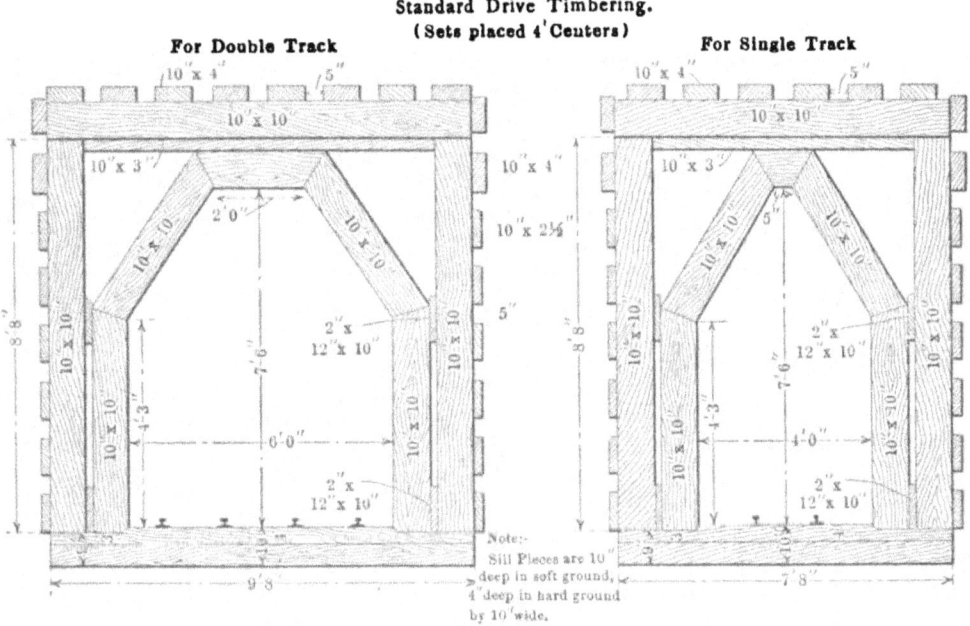

FIG. 102

acting catch holds it steady, while it is lowered and discharges into the bin. The skip is again hoisted and run out along the cable and again lowered into the cut. In the meantime a second skip has been filled, and is attached, hoisted and discharged as described.

The skips are about 16 in. deep, 4 ft. wide and 5 ft. long. They are suspended by four chains, one at each corner, the back two being fixed to the skip, while the front ones are fastened by hooks, which are undone to release the load. The skips are used for many purposes. I have seen a workman with a broken leg hauled to the surface and deposited safely, thus saving a climb up the banks. The sides of the cut are made with a slope of one-half to one and three-quarters to one, and would meet at a depth of 250 ft.; but owing to the frequent slips of the sides they are often irregular. Some very heavy blasting is done in these cuts, which shakes the ground for a great distance, but great masses of ore are removed.

The sets as used in the double-track drives, where horses are used underground, are framed as shown in Fig. 102. They are designed to withstand heavy pressure, and are placed at intervals varying with the nature of the ground they have to hold, but are generally 4 ft. to 5 ft. between centers.

Chapter XXV

MINING AT THE HOMESTAKE, LEAD, SOUTH DAKOTA

The Homestake mine is without exception the largest gold mine in the world. The great property is operated through six shafts, though a large amount of ore is still mined in open cuts by the usual methods, the broken ore passing downward through mill-holes to chutes on levels below, where it is drawn out into cars, trammed to one of the shafts, and hoisted to the surface. Until recently all hoisting was done by means of cages, but lately skips have been introduced at some of the shafts. With the exception of the Ellison shaft, all of the several shafts of the Homestake property are of the usual rectangular, 3-compartment type. The Ellison shaft, however, differs materially from the others in its dimensions, having two hoisting compartments, each 5×10 ft., and a pump and manway compartment 6×10 ft. clear. Figs. 103 and 104 show the method of timbering in this shaft. Timbers were loaded on flat cars, run on to the cages and lowered to the level where they were required. This saved much time in handling timbers and was found superior to the usual method of standing them on end and lashing them to the cage frame.

The stations at the shafts of the Homestake were all of liberal size, and were cut from 12 to 16 ft. high. Iron plates were laid on the floors, so that cars could be shifted about anywhere between the track ends and the shaft, but the sheets were later taken up and the rails were carried directly up to the shaft, opposite each hoisting compartment. The tracks in the stations were laid with turnouts, so that cars could be sent from both main tracks to either cage in the shaft. Fig. 105 shows the arrangement of tracks at stations of the Ellison shaft.

The main drifts in this property are generally of a single type, illustrated by Fig. 106. All the drifts are driven with

230 TIMBERING AND MINING

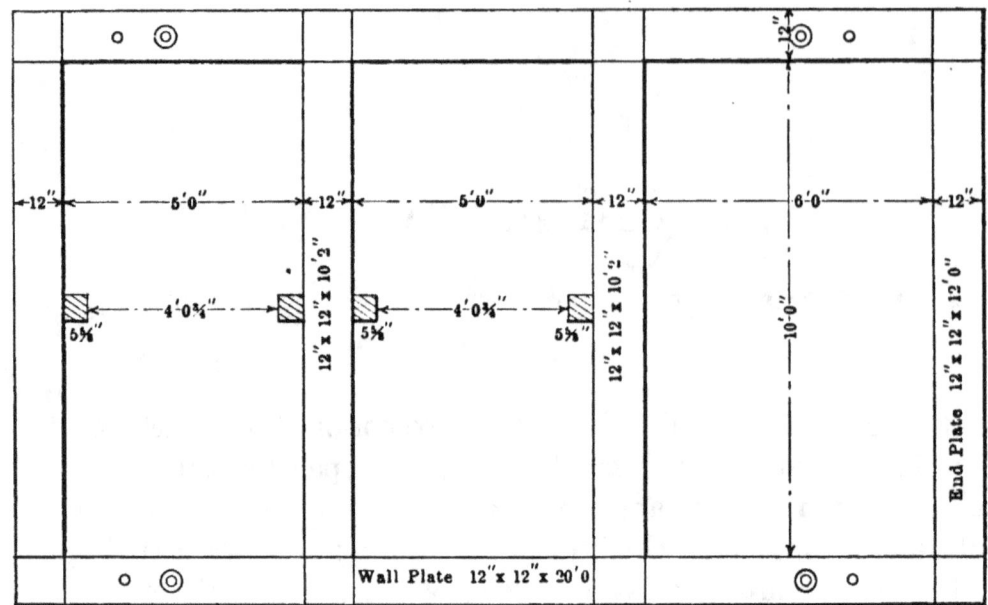

Fig. 103. — Plan of Ellison Shaft

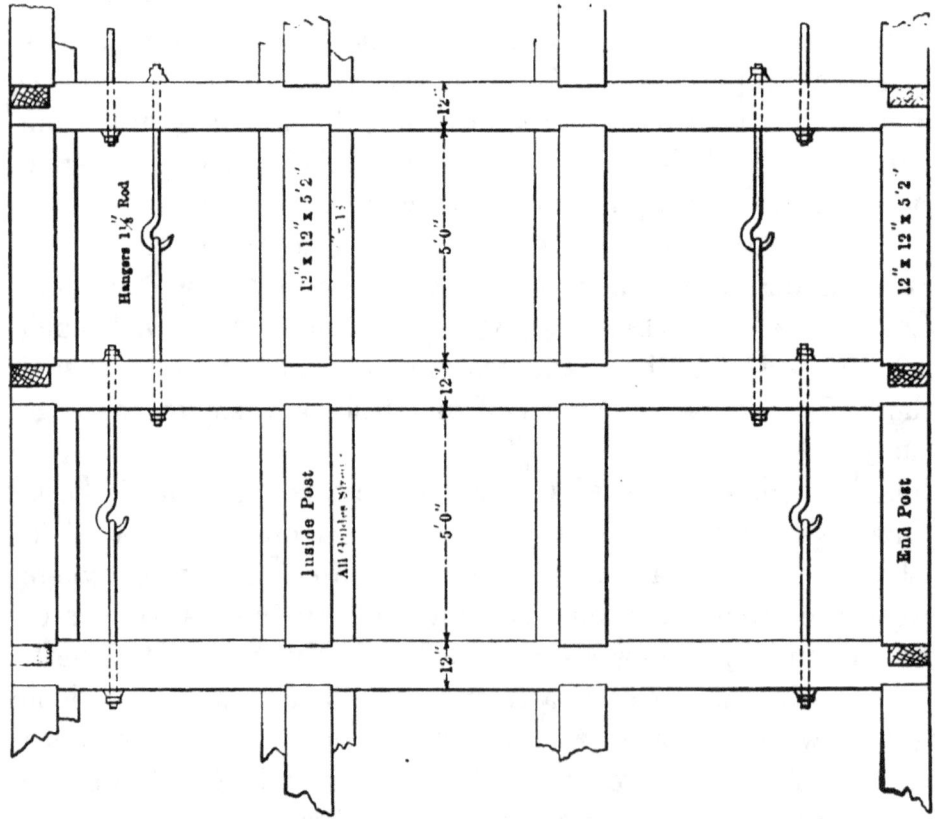

Fig. 104. — Longitudinal Vertical Section Ellison Shaft

machine drills. Raises are made for various purposes, ventilation, ore-storage bins, transferring ore or waste from one level to another, or to provide an opening through which waste may be dumped into a finished stope. Permanent raises for ventilation, ore bins, etc., are located in the country rock far enough away from the ore body to be undisturbed by mining operations. The

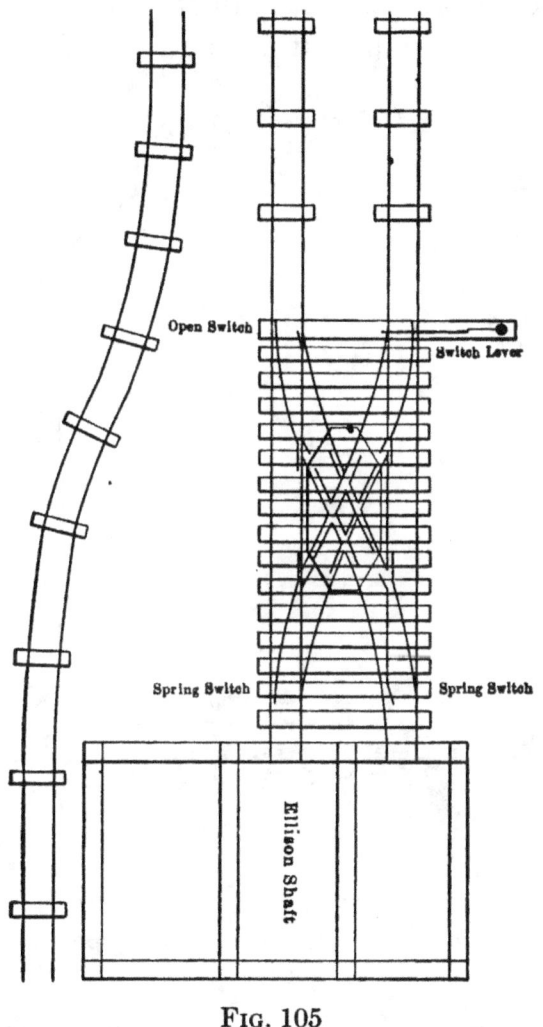

Fig. 105

purpose of the waste raise is to provide a storage for waste taken from dead work when not immediately needed in the stopes, and these are so arranged that waste may be drawn out or dumped in at any level. The main waste raises are connected with the surface, and the porphyry, which forms a cap overlying the ore-bearing formation, is drawn through them and used for filling. Fig. 107 shows the arrangement of one of these continuous raises, which may be used for waste or ore. The ore raises do not, as a rule, connect with the surface.

On one or several levels, if convenient, cross-cuts are driven to the point from which it is desired to make a raise, care being taken to have the cross-cuts on the different levels alternate on each side of a vertical plane. When the cross-cut has been driven far enough beyond the position of the raise to provide a passing track for empty cars, the drift is widened; four sets of regular stope timbers are put in, in the form of a square, and an

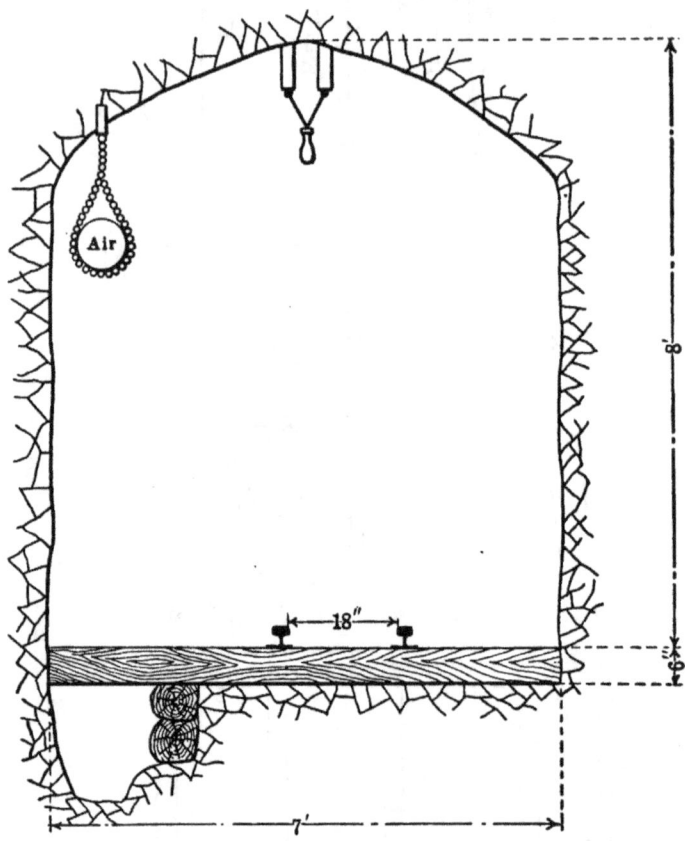

Fig. 106. — Cross-Section of Drift Showing Air pipe, Drain and Electric Light

ordinary board chute built in one set. Above these timbers the raise is gradually drawn in to a 6 × 6-ft. raise. From this point the raise is carried up in the usual manner. Sprags across the raise, about 5 ft. apart, are used to sustain the working platform and serve as a ladder way. The smaller size machine-drills are used, and are hoisted and lowered by rope and pulley. The raise on the level below is located about 15 ft. to one side or the other, and is cut straight to the upper level. After connection has been made, an inclined by-pass is made to connect with the upper raise above the timbers. Another small inclined raise connects the lower raise with the cross-cut, at the top of which are located the grizzlies. The bars forming the grizzlies are spaced about one foot in the clear so that no large rock can get into the raise. The successful operation of these continuous raises depends in a great measure upon the grizzlies. The by-pass is closed by a gate made of steel plate, sliding in cast-iron grooves fastened to upright timbers, and operated by rack and pinion. A similar gate is used in the main ore bins under the

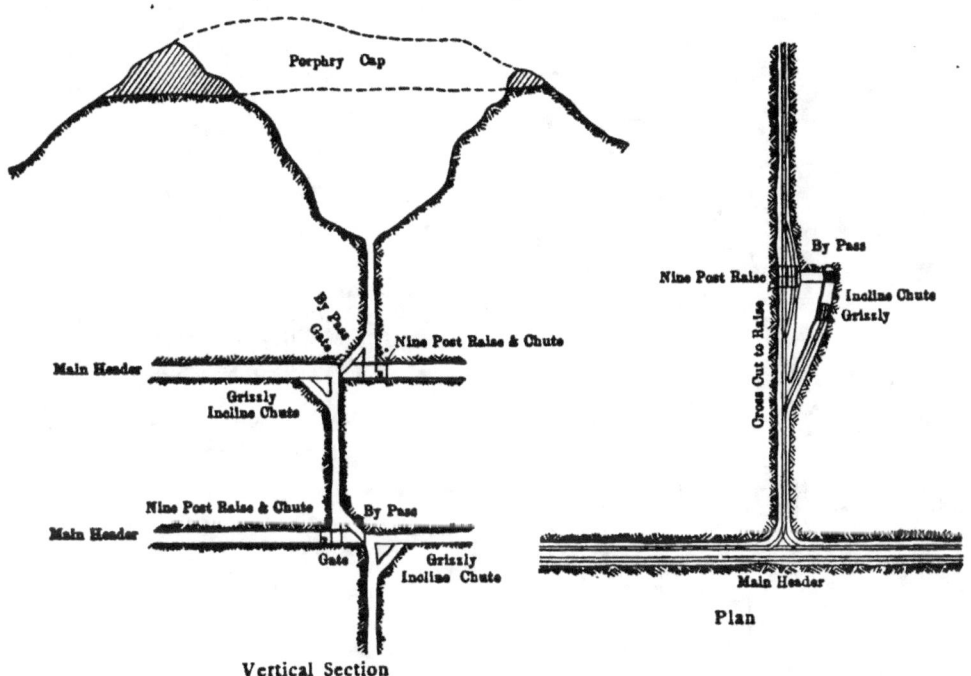

FIG. 107. — Continuous Raise

crushers. It works easily, and can be operated gradually, which prevents rushes of rock. Similar arrangements are made on the other levels. As soon as one raise is made it can be put into service and the raise from the lower levels finished when needed.

In making an ore-storage bin, the timber sets are carried up six or seven posts high. When connection is made with the level above, the timbers are removed, with the exception of the sill floor in which the chutes are located. The by-pass and small raise to the grizzlies are located as in the waste raise. Ingersoll A-32 machines are usually used in making raises. One miner and helper will make a 6 × 6-ft. raise 100 ft. high in ordinary ground in sixty shifts; provided they are not required to tram the rock. Blasts are fired by ordinary caps and fuse. When a stope has been worked nine or ten floors high, raises are put up in convenient places to the level above, through which the filling is dumped. These raises are all in ore, and consequently pay their own way.

When a raise is to be made near the face of a long tunnel where the air is bad, some artificial means of ventilation must

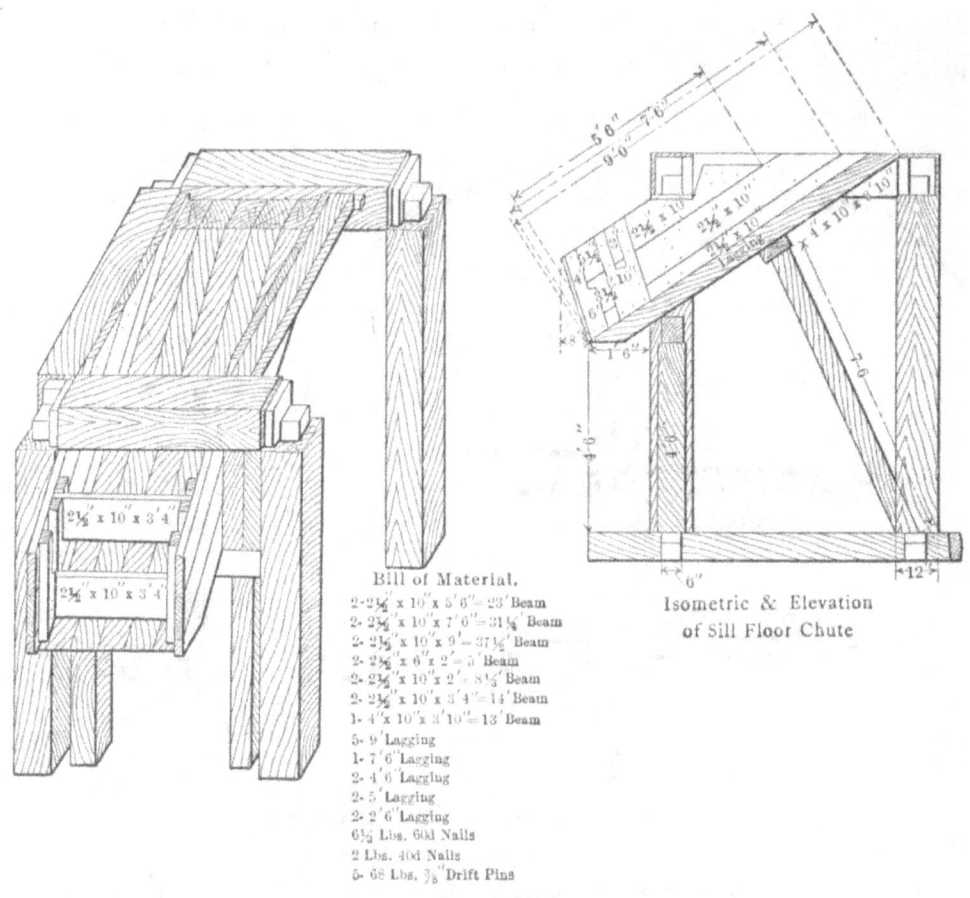

Fig. 108

be provided. A device introduced here has proven successful. A 6 or 8 in. light iron pipe is laid from the entrance of the drift to the foot of the raise. Near the entrance a small ½-in. pipe is tapped into the main air pipe and brought down to and into the large pipe, with the end which projects into the pipe turned out. It has been found that a small amount of air under a pressure of from 75 to 80 lb. will effectually clear the raise in a few minutes. This device is used also in running long drifts. An exhaust fan would possibly be a more economical machine so far as power is concerned, but the first cost would be greater, and the air is used only when the machine drill is not in service. All chutes are of the ordinary type, with bottoms made of lagging and sides of 2-in. plank. The rock is held back by two boards, one higher than, but not directly above, the other. Fig. 108 shows the method of building these chutes.

The Homestake mills, six in number, contain 1000 heavy stamps, the duty of each stamp being about 4 tons per 24 hours. These mills are now crushing annually over 1,400,000 tons of ore — more than 4,000 tons daily. About 80 per cent. of this vast tonnage is taken from underground stopes, the remainder from the open cuts. The ore obtained from the open cuts is broken down from the sides into openings that connect with the regular levels of the mine. These openings or raises are provided with chutes on the different levels, from which the cars are loaded. The loaded cars are made up into trains of from four to eight cars, according to the grade of track, and are drawn by horses to a central point and long trains are then hauled to the shaft by compressed air motors.

The ore taken from open cuts is mined very cheaply. Two miners whose wages are $3.50 per day will break, on the average, 200 tons in one shift. Two men are employed at the chutes, one to break the rock at the grizzlies and one to load the car. The grizzlies are located two or three floors above the chute. Ore and waste are blasted down together, and the waste is sorted at the chute and used for filling.

In one of the open cuts now being worked, the process of mining is much like the work of making a very deep cut through a hill some 300 ft. high. Tracks are laid from the crushers to one end of the cut, and the rock is blasted down, loaded into cars, and trammed directly to the crushers. The pillars and backs of old stopes which were left in place in the older workings of the mine are removed by a system of draw raises.

It is very important to know the exact location of these pillars and roofs before starting this system. If the maps and stope records have been accurately kept this becomes an easy matter; otherwise the memory of some old employe must be depended upon to give the necessary information. Having located the ore, a nine-post raise, which consists of four regular stope sets arranged in a square, is put up either on the foot-wall or on the hanging-wall side, preferably the foot-wall. This raise is carried up a sufficient height to reach the ore above the waste filling. Grizzlies are put in on the floor next to the top and the sides of the raise are carefully lagged to protect the men. The run of ore is then started by a blast or by barring. A man stationed at the grizzly breaks the rock so that it will pass through into the chute which is located on the sill floor.

The timber required to make a nine-post raise four floors high consists of 36 posts, 24 caps, 24 ties, about 150 lagging, 4 stope ladders, and one ordinary board chute. The grizzlies are made by placing two or three 12-in. timbers, spaced 12 in. apart, over one set of timbers, and protecting these by pieces of sheet iron curved to fit. The posts on the top floor may be protected in a similar manner from the blasts and from running rocks. Fig. 108 illustrates a nine-post raise, and the method of working a caved stope.

Sixteen thousand tons of ore were taken from one raise in six months; and this was done in a place where vain attempts had been made to reach the ore by carrying up timbered stopes through the old fill. Should the ore be too solid to run, the raise will serve as a manway to a timbered stope started on the top of the fill. Whenever a run of waste is encountered, it is drawn down and used for filling in other parts of the mine, and the ore from upper levels will follow the waste down.

In former years, all stopes in the Homestake and Highland mines were timbered by the square-set method. An enormous amount of timber was required in the excavations, and the work of handling the timber through the shafts made it difficult to hoist sufficient ore to supply the stamps. The Nevada square-set system was employed wherever square-sets were used in these mines. The system was employed as soon as underground mining began on the lode in 1878. No unusual features were introduced, and as filling was neglected in those early days, numerous disastrous caves occurred in nearly every large mine on the lode. The cave in the Caledonia mine, one of the group, was a particularly interesting one. A cross-cut adit over 800 ft. in length was run to the main ore body, which was over 200 ft. wide measured horizontally. A vertical shaft was sunk from the adit level in the center of this mass of ore, reaching the foot-wall in a depth of about 100 ft. A pillar of ore 30 ft. square was left around the shaft when stoping began on this level and the next below it, but it was insufficient, and the great weight of the

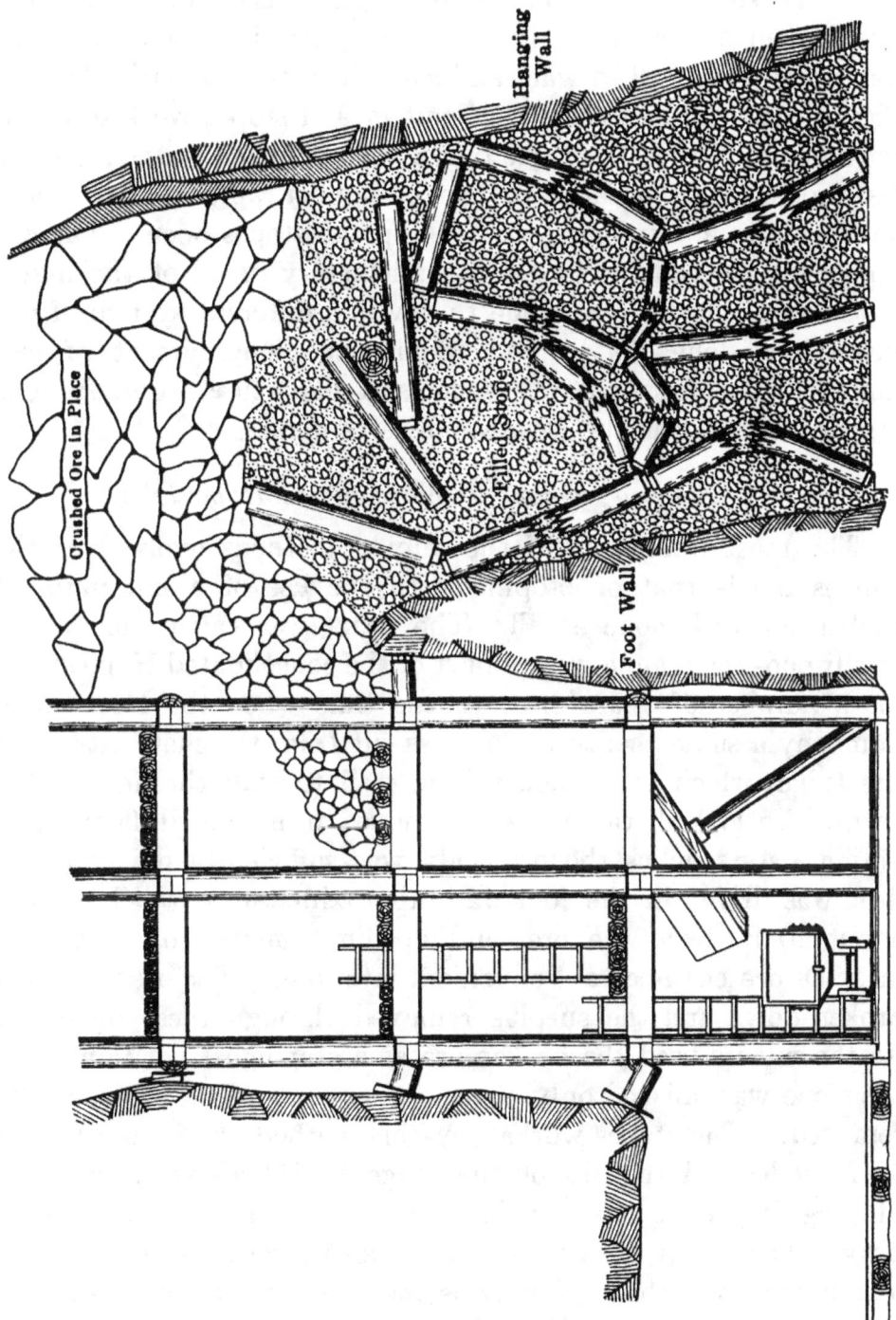

Fig. 109. — Nine Post Raise to Work Back of Caved Stope

superincumbent ore and hanging country caused the great timbers of the square-set to jack-knife, and the upper workings — machinery, timbers and all — plunged in a chaotic mass into the stope below.

In the Homestake mine, on the upper levels, where the ledge was broken and comparatively narrow, the sill-floor excavation was made from wall to wall, and sometimes for the entire length. The sill-floor timbers were then put in and stopes worked where convenient. When the ledge began to widen this method proved disastrous, and thousands of feet of lumber were used to make bulkheads in a vain endeavor to keep the stopes open. The old square-set method proving so unsatisfactory in all of the mines of the lode, efforts were made to devise a different system of ore extraction — one that would require the employment of less timber. The first notably successful departure from the old practice was that in the Deadwood-Terra mine.

Stoping without Timbers at the Homestake

The most important of the mining methods evolved at the Homestake is that of stoping with the use of a minimum of timbers or with none at all. The latter was carried on in the Deadwood-Terra mine, now a part of the consolidated Homestake, for a number of years prior to the time when the Homestake Company assumed control. A cross-cut from the shaft intersects the ledge, which was opened from wall to wall throughout the entire length. All the ore was removed from the sill-floor excavation. As soon as the ore body was sufficiently developed, a drift was driven in the foot-wall, approximately parallel to, and about 20 ft. from the ore, and openings made from the drift into the ore chamber at convenient intervals. The ore was then broken down and the surplus removed through these openings. As the miners used the broken ore as a staging to work on, until the stope was finished only about 40 per cent. of the ore could be removed. The stopes worked by this method are from 30 ft. to 50 ft. wide, and the dip of the ledge is 75° above horizontal. Stoping may be carried on in several levels at the same time, provided sufficient back is left in between levels. When the level above is finished, this back may be caved, and all the rock removed on the level below. No timber is required other than a few lagging for staging, but the method is not applicable to wide ledges. Fig. 110 shows a cross-section of several levels and a part plan of one level, showing the method of operation. The sketch of vertical section fails to show the raises necessary for manway and for ventilation. This is shown, however, in the plan.

A new method was inaugurated by W. S. O'Brien on the 600-ft. level of the Homestake mine. The plan is as follows: A cross-cut is driven through the ledge from the central shaft, which is called the main cross-cut. The ore body is then developed by driving a 24-ft. drift along the foot-wall. From this drift stopes 60 ft. wide are opened across the ledge (500 ft.), with 60-ft. pillars between each stope. Beginning with the main cross-cut, a pillar is left on each side, then a 60-ft. room, a 60-ft. pillar, and so on

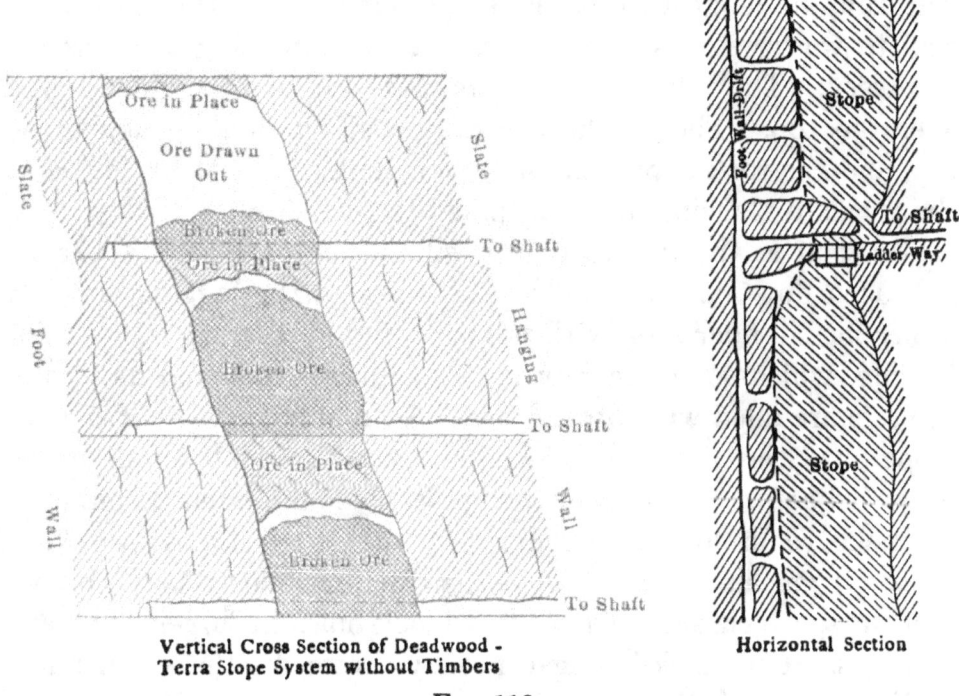

Vertical Cross Section of Deadwood - Terra Stope System without Timbers

Horizontal Section

FIG. 110

to the end of the ore body. These rooms and pillars are numbered north and south of the main cross-cut. No. 3 stope north would be the third stope north of the cross-cut, and No 4 stope south the fourth stope south of the cross-cut. Some difficulty was experienced in keeping the sides of the rooms straight while the sill floor was being opened. To overcome this sills were laid in the foot-wall drift to lines given by the surveyor, and the miners took their lines from these. Each stope contains eleven lines of sills. When the system once became established, no difficulty was experienced in keeping the room of uniform width and the sides comparatively straight. When a stope has been worked and filled, the pillar may be attacked, the lagging on the sides keeping the filling in place so that all the ore in the pillar may be removed. In these stopes the square-set system of timbering was at first employed.

However satisfactory this method of blocking out the ore body may be, it did not solve the timber problem. It was finally decided to try stoping in these rooms without timbers. As the result of the experiments the level is opened by the room or block method, and sills are laid in the rooms the same as for timbered stopes. When the sills are in, three lines of track are laid, running lengthwise of the stope but crossing the ledge from wall to wall, with as many cross tracks connecting them as are necessary. The sill-floor sets are put up, and lagging placed over the top. The tracks are protected by double lagging on top, and the rock is prevented from running in at the sides on to the tracks by lagging or slabs spiked to the posts.

As soon as the timber is in position stoping begins. The ore is broken down and allowed to fall through the lagging, entirely filling the sill-floor sets, with the exception of the car ways. The lagging, which serves merely as a staging, is removed as fast as the sets are filled with broken ore. No rock is removed from the stope until this filling is finished. When the next cut or breast is carried across the back of the stope, some ore must be removed to make room for the miners. In the large stopes two D-24 Ingersoll machines are employed, with from one to two "baby" machines, which are used to drill block-holes in larger slabs and boulders. For block-holing generally, however, the small pneumatic-hammer drills are used to great advantage. This machine will quickly drill holes from 6 to 12 in. deep, and has proven very successful in block-holing large boulders in the open stopes.

As there are no timbers to break, no limit is placed on the miner as to the amount of rock he may bring down at one blast. The stope should be finished as quickly as possible, so that the broken rock may all be removed if needed. Consequently, large slabs of ore are blasted down, and these must be broken up to car size, either on top of the pile or on the sill floor, as it is drawn down by the shovelers. On account of the uneven size of the rocks, chutes are not generally used in these stopes, muckers shoveling the ore into cars from the level of the track, there being

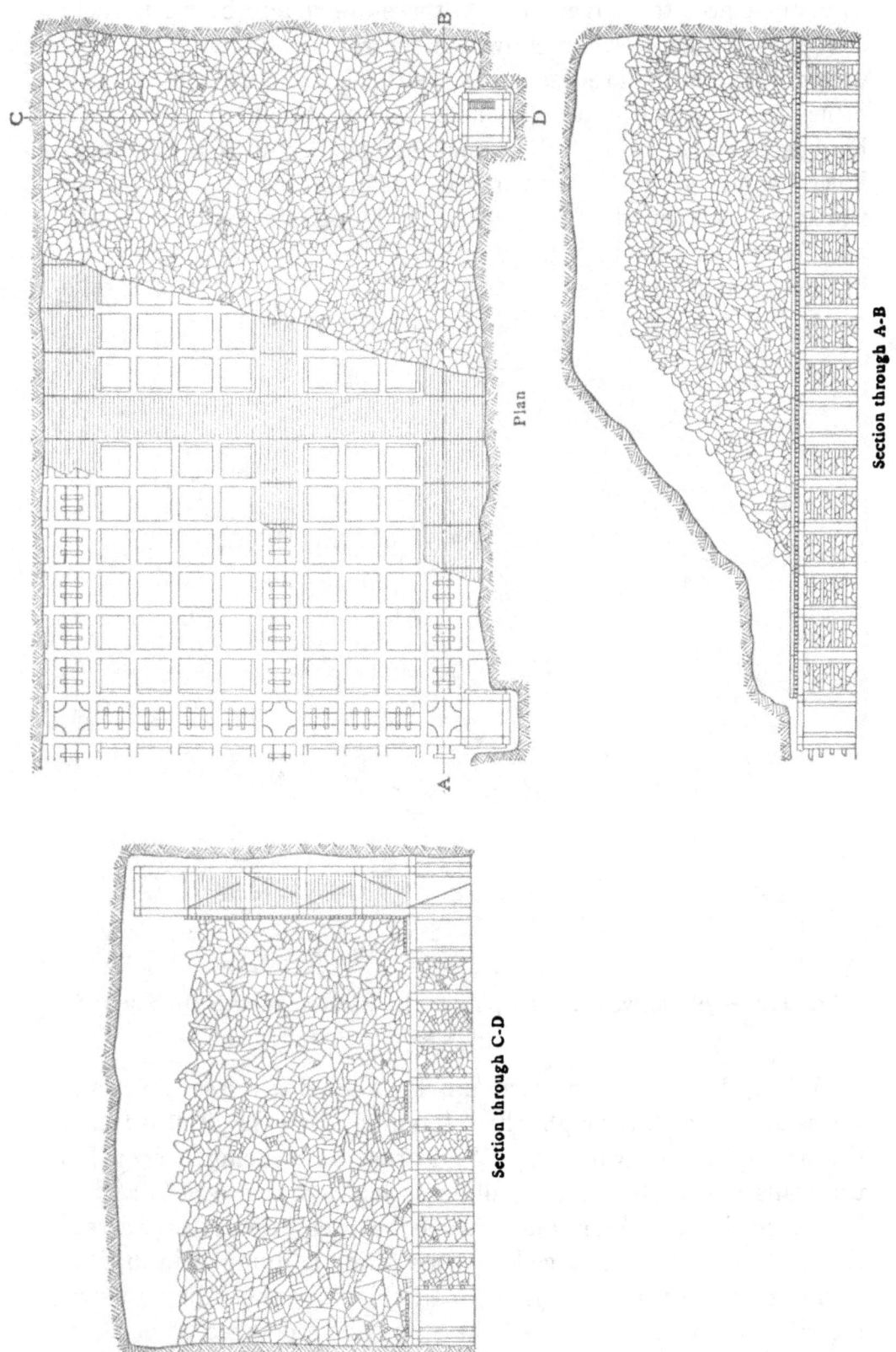

Fig. 111

as many places to shovel from as there are spaces between posts along the tracks. However, where the rock is soft and where it breaks fine, chutes are used to advantage. Should a large rock come down which the shovelers cannot break with a rock hammer, the car is moved to another opening until the "blockholer" comes around. Two or three regular sets on each side of the stope are carried up as fast as the stope is worked, in which are placed the ladders and air pipes. These open sets also assist in ventilating the stope. Fig. 111 illustrates the Homestake method of stoping with the minimum of timbers.

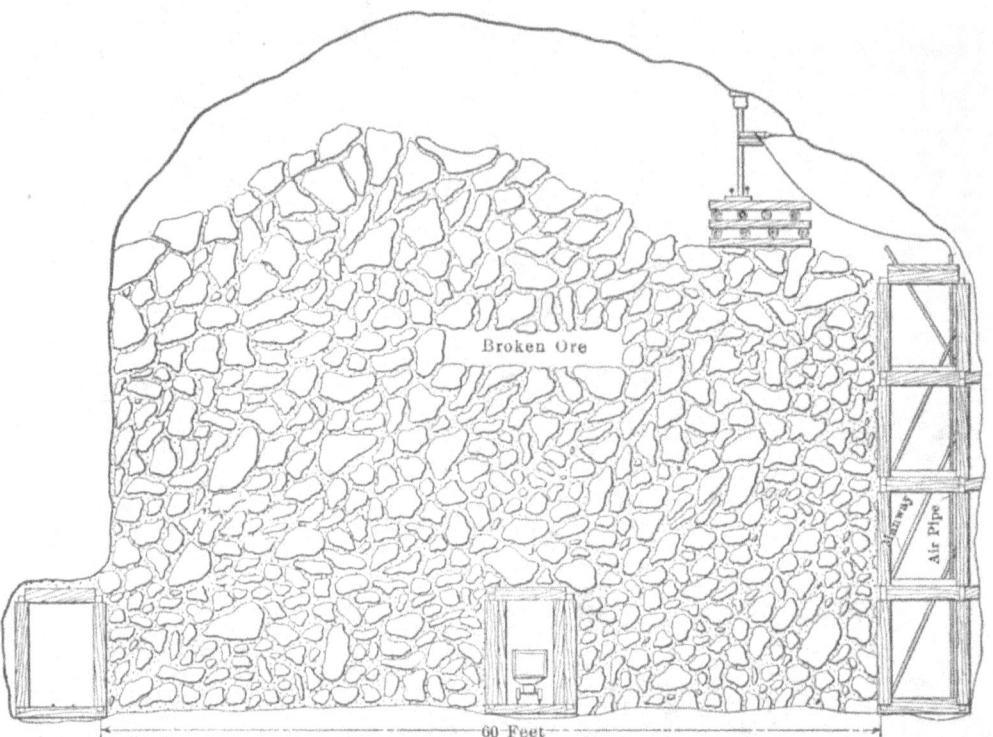

Fig. 112. — Block System of Stoping at the Homestake with Minimum of Timbers

When the stope is worked up 80 or 85 ft., raises are made to the level above, through which the filling is to be dumped, and the ore is then drawn out. While the ore is being drawn out the walls and roofs are carefully watched and all loose material is dressed down. When one end has been emptied of ore, a section of the sill floor is lagged and the filling is dumped in until it begins to run over the lagging. In this way the filling follows the shovelers and the walls of the stope are supported at one end by the ore and at the other by the waste. When small ore bodies are worked by this method no pillars are left, but when one section is worked to a sufficient height, another section is started at one end and the ore is left in until the entire body is worked. As only a small per cent. of the ore can be removed before the stope is finished, there is of necessity a large reserve always on hand, which allows the mine to lay off miners whenever desirable. The broken ore reserve in the mine is over 1,000,000 tons.

Another plan of working these stopes has been evolved from the experience gained by employment of the method above described, by means of which at least 50 per cent. of the sill-floor timbers is saved. Instead of the two outside tracks being laid in the stope, a drift is cut under the edge of each pillar of sufficient height and width to receive one line of regular stope sets, and tracks are laid in these. The timbers are put in place, and the ore is broken down until the sill chamber is partly filled, the broken rock holding the timber sets in place. A drift is run through the broken rock in the center of the stope, in which another line of sets is put up. No sills are laid between the rows, and only six rows of posts are required, whereas eleven rows are required in the other stopes. The tracks are laid in these open sets and the two under the pillars may be used again when the pillar is being taken out.

The operation of these stopes has brought out another change in the general working plan of the mine which will make a material saving in the cost of development. Timbered stopes cannot be worked economically more than 80 or 90 ft. high, as the timber begins to crush of its own weight, while the untimbered stopes can be carried up 150 ft. as well as 100 ft. As a result, the distance between new levels will be made 150 ft., which will effect a saving of one-third in development work, and one-third more ore can be recovered from the stopes with the same sill-floor work.

Fig. 111 illustrates the Homestake methods of stoping with, and Fig. 112 the method without, sill-floor sets. That employing the timber sets on the sill floor is similar to the method in use at Broken Hill, New South Wales, described and illustrated elsewhere.

Usefulness of Mine Models

A good model of the workings of a mine is of great value in laying out work and in studying mine methods. With a properly constructed model, the superintendent and foremen can advantageously study out improved methods and note the probable effect upon existing workings and on those yet to be made. There is in the office of the Homestake Company an elaborate model of a timbered stope. The model is about 5 ft. square and has been carefully made, showing every detail of the work. Not a stick of timber used in actual practice is absent in the model, and by a careful study of this miniature stope inportant changes in methods have been worked out. The management of the Treadwell mine, in Alaska, employs a model of that property, made from plaster of Paris, for similar and other purposes, the model being sawed into sections. The managers of other properties make use of mine models, and they are always found useful, as they make it possible to present the broader problems involved within a space immediately under the eye. Maps, of course, have similar advantages, but a properly constructed model is better for the purpose of studying practical mining problems than the best map.

The following table, prepared by B. C. Yates, engineer for the Homestake Company, shows the comparative cost of mining with the old square-set method and by the later method, where relatively little timber is used:

Timbered Stope

Name of Piece	Number of Pieces	Lineal Feet or Feet Board Measure	Cost of Material	Labor, Sawing and Framing	Total
Sill-floor posts ..	421	3,650	$474.50	$96.83	$ 571.33
Upper-floor posts	2,077	16,616	2,160.08	477.71	2,637.79
Caps............	2,410	13,255	1,723.15	506.10	2,229.25
Ties.............	2,261	12,435	1,616.55	474.81	2,091.36
Sills — 203 long, 382 short	....	4,537	226.85	22.69	249.54
Lagging	13,020	75,906	3,795.30	379.53	4,174.83
Lagging strips ..	2,410	4,025	64.82	30.00	94.82
Wedges	2,352	784	13.33	11.76	25.09
47 sill-floor chutes — complete ..	....	....	311.68	16.25	327.93
215 upper-floor bins — complete	....	....	786.22	37.90	824.12
Ladders	14	117	1 99	3.50	5.49
Labor placing timbers and chutes........	....	....	....	....	4.745.00
Breakage (10% of lagging, 5% posts, caps and ties	....	....	....	....	793.97
Totals	....	....	$11,174.47	$2,057.08	$18,770.52

Stope Worked by Homestake Method

Sill-floor posts ..	421	3,650	474.50	96.83	571.33
Caps...........	410	2,250	293.15	86.10	379.25
Ties............	381	2,095	272.35	80.01	352.36
Sills, long	203	2,436	121.80	12.18	133.98
Sills, short	382	2,101	105.05	10.50	115.55
Lagging	1,752	10,214	510.70	51.07	561.77
Lagging to protect track.....	764	4,454	222.70	22.27	244.97
Relief lagging ...	1,684	13,472	673.60	67.36	740.96
Wedges	200	66	1.12	1.00	2.12

Manways

Upper-floor posts	96	768	99.84	22.08	121.92
Caps	48	264	34.32	10.08	44.40
Ties	48	264	34.32	10.08	44.40
Lagging, floors	96	560	28.00	2.80	30.80
Lagging, sides	720	4,197	209.85	20.98	230.83
Drift pins	1,440	457 lbs.	22.85		22.85
Ladders	28	235	4.00	7.00	11.00
Labor standing sill-floor timbers					758.16
Totals			$3,108.15	$500.34	$4,366.65

Seventy-three thousand tons were taken from this stope. $18,770.52 ÷ 73,000 = $0.257 per ton by former method; $4366.65 ÷ 73,000 = $0.060 per ton by Homestake method; $0.257 — $0.060 = $0.197 saving per ton.

The ore of the lower levels of the Homestake mine is hornblendic schist containing considerable finely disseminated iron sulphide. This rock weighs 200 lb. per cubic foot, therefore but 10 cu. ft. (instead of 13 ft., as usual) are required to weigh one ton.

Chapter XXVI

WORKING DANGEROUS GROUND IN THE KIMBERLEY DIAMOND MINES

All who are familiar with the diamond mines of South Africa are aware that in the early years of mining in that field the mines were worked as open pits, the largest of which covered many acres. The pits varied somewhat in depth, but in the later years were 500 ft. below the rim at the deepest place. The diamondiferous ground was worked up to the barren country rock which surrounded the crater, leaving the great open cut with vertical walls. Rock caved from the rim as the work proceeded, and as the central portion of the pit grew deeper, the frequency and danger of caves of rim rock increased, thousands of tons falling at a time, until finally open-pit mining was discontinued altogether.

Vertical shafts were then sunk in the country rock at a distance of several hundred feet from the edge of the pit. Levels were driven from the shafts to the diamond-bearing ground some distance beneath the lowest portion of the open pit, and the ground removed by stoping. The removal of that portion just underneath the débris caved from the rim is attended with great danger, but a method of recovering the ground has been devised which makes it fairly safe.

Gardner F. Williams thus describes the method of mining introduced by him at Kimberley, in his splendid work, "The Diamond Mines of South Africa": "Instead of attempting to withstand, even for a time, the pressure of the superincumbent mass of broken reef (this is barren rock that fell into the great open pit from the walls of the crater, thousands of tons of which overlaid the diamond-bearing ground in the huge chimney or deposit), the new system introduced was a caving in and a filling of the excavations after precious blue ground had been extracted. When numerous small tunnels had been driven to the margin of the mine, that is, to the point where they reached the sides of the crater, the blue ground was stoped on both sides of, and above, each tunnel until a chamber was formed extending along the face of the rock (wall) for 100 ft. or more, with an average width of 20 ft., and about 20 ft. high. The roof of the chamber or gallery was then blasted down or allowed to break down by the pressure of the overlying mass of broken diamond-bearing ground or (the barren) débris.

"In the early stages of underground mining there was an enormous amount of diamond-bearing ground which had been left behind when open mining was discontinued, and which had been crushed either by the moving sides of the immense opening or by the collapse of the underground pillars when mined by the old system (of pillar and room). It happened frequently, after breaking through to the loose ground above, that clean diamond-bearing ground would run down as fast as it was removed for weeks or months at a time. The galleries would at times become blocked with large pieces of blue ground, which had to be blasted, and then a further run of blue ground would follow. When the blue ground was worked back toward the center of the crater, larger boulders or fragments of basalt which had come down through the loose reef from the surface would be met with. This system of working would be continued until reef alone came down, the waste or reef removed being sent to the surface by itself and piled on the waste dump. It formed only an inconsiderable proportion (one to four per cent.) of the total output. When the roof caved in, the gallery was nearly full of blue ground. Only a part of this ground was removed by the men working on that level, the miners preferring to take it out on the next level below. This process of mining was repeated from level to level until finally there was no more loose ground to be recovered. The cost of extracting blue ground while loose ground existed was very low.

"Now all this is changed, and the plan of opening new levels has altered somewhat, but the system remains the same. When the underground work had reached a depth of 800 ft. or more, a new danger appeared. The huge open mines are filled with débris from the sides, caused by the removal of the diamond-bearing ground by open quarrying to depths varying from 200 to 500 ft. As the supports were removed the sides caved and filled the open mine. This débris was composed of the surface red soil, decomposed basalt, and friable shale, which extended

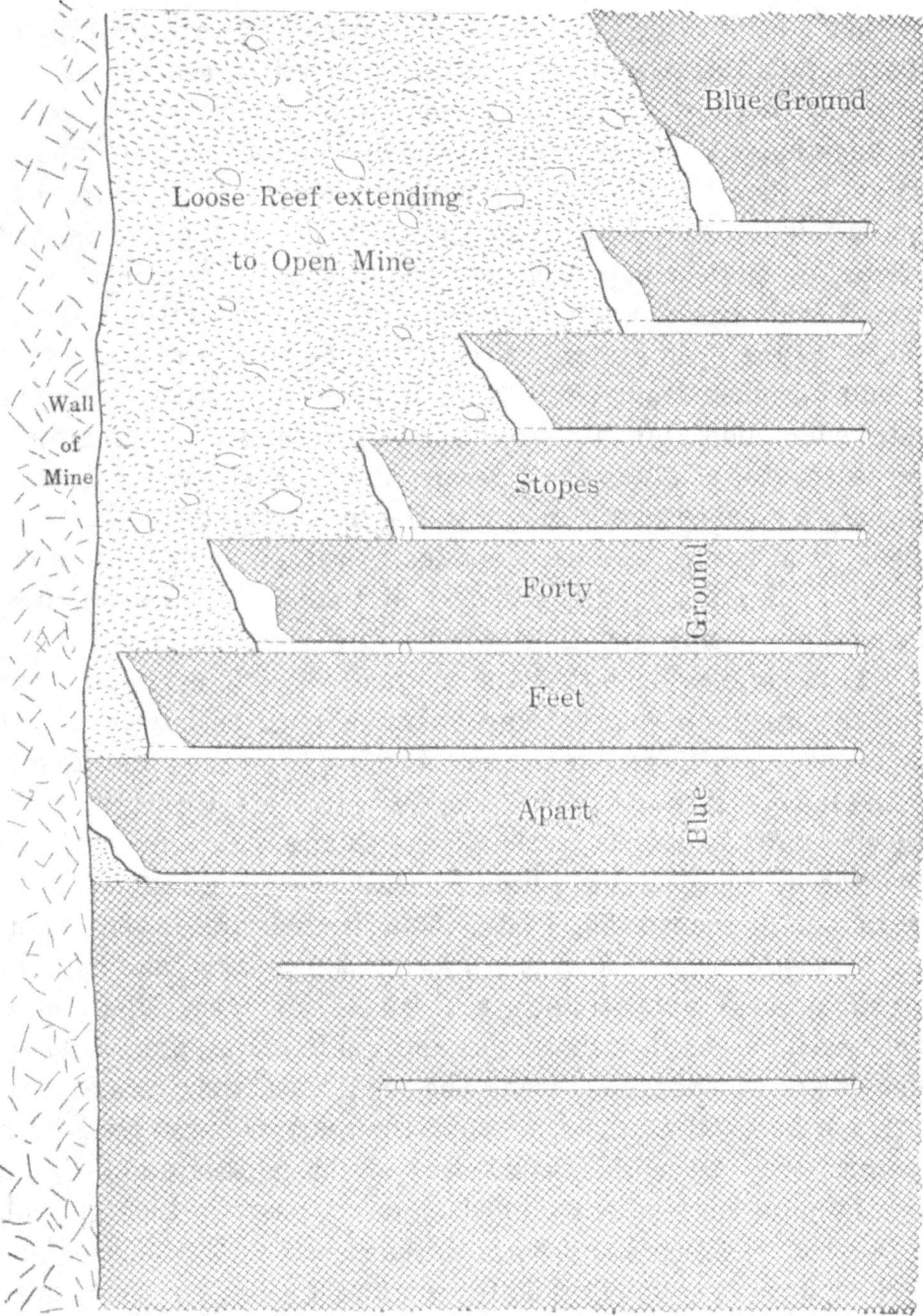

Fig. 113. — Vertical Section Kimberley Diamond Mine

from the surface down to a depth of about 300 ft. In addition to the débris from the surrounding rocks, there were huge masses of 'floating shale,' resembling indurated blue clay more than shale. Large heaps of yellow ground and tailings, which the early diggers had deposited near the margin of the mines, and west-end yellow ground, contributed to the mud-making material. The black shale which surrounds the mines disintegrates rapidly when it falls into them. It contains a small percentage of carbonaceous matter, and a large amount of iron pyrites. When the huge masses of shale fell into the open mine, they frequently ignited, either by friction or, more probably, by spontaneous combustion, as they have been known to do on the dumps, and burned for months and even years at a time. These masses of burned shale become soft clay and form a part of the mixture which fills the open crater. This débris moves down as the blue ground is mined from beneath it, and becomes mixed with the water which flows into the open mine from the surrounding rock, and with storm water, and forms mud. This overlying mud becomes a menace to the men working in the levels below. Frequent 'mud rushes' occurred suddenly, without the least warning, and filled up hundreds of feet of tunnel in a few minutes, the workmen being sometimes caught in the moving mass.

"It became evident that the method of working was dangerous, the men sometimes being, when a mud rush took place, either shut in or buried in the mud coming from the opposite end of the mine. It was decided, therefore, to work the mines from one side only, and to have the offsets to the rock connected one with the other at as few points as would be consistent with the ventilation of the working faces. Main tunnels are driven (about 100 ft. apart) across the crater upon its longer axis, and at right angles to these small tunnels are driven out every 30 ft. until they reach the hard rock on the south side of the mine. These tunnels are widened, first along the rock, until they connect one with another, and at the same time the roofs or backs are stoped up until they are within a few feet of the loose ground above, thus forming long galleries, filled more or less with blue ground, upon which the men stand when drilling holes in the backs. The working levels were at first 30 ft. apart vertically, but for greater economy the distance was soon changed to 40 ft.

"The broken blue ground in the galleries is taken out, as a rule, before there are any signs of the roof giving away. At times this is impossible, and the roofs cave upon the broken ground, and the blue ground is covered with (barren) reef." Figs. 113

and 114 show the arrangement of cross-cuts and drifts and the manner of stoping under the conditions described.

"As the roof caves or is blasted down, the blue ground is removed, and the loose reef lying above it comes down and fills

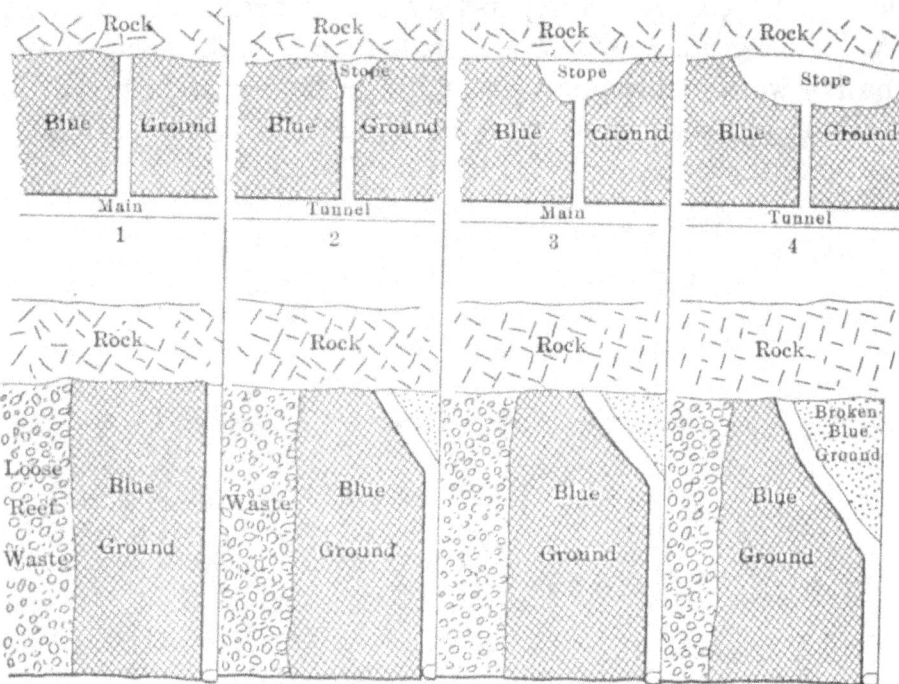

Plan and Vertical Section in Stope Kimberley Diamond Mine.

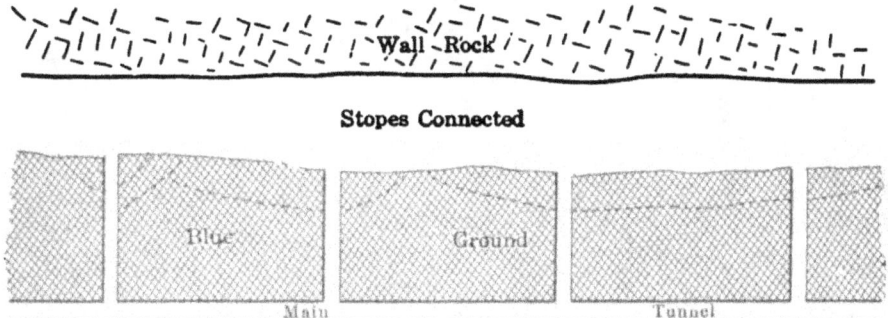

Fig. 114

the gallery. Tunnels are often driven through the loose reef, and the blue ground which has been cut off and buried by débris is taken out; but it is sometimes left for those working the next level below to extract.

"After the first 'cut' near the rock is worked out, another cut is made, and in this manner the various levels are worked back, the upper level in advance of the one below, forming terraces as shown in the accompanying sketch. The galleries are not supported in any way with timbers, but all tunnels in soft blue ground are timbered with sets of two props and a cap of round timber, and are covered with inch and a half lagging. Soft blue ground is drilled with jumper-drills sharpened at both ends. In hard blue ground drills and single-hand hammers are used. The native workers become very skilful in both methods of drilling, and do quite as much work as white men would do under similar conditions."

Chapter XXVII

THE DELPRAT METHOD OF STOPING WITHOUT TIMBERS

G. D. Delprat, in "Transactions of the American Institute of Mining Egnineers," Vol. XXI, describes a method of stoping a large body of ore in Spain in which no timber was employed. The ore body was 500 ft. long and from 20 to 75 ft. wide, the average being 32 ft. The lode is nearly vertical — about 75°. A main extraction shaft was sunk in the hanging-wall, some distance from the vein at the surface, and approaching it in depth. A pumping shaft was also sunk in the hanging, some distance from the main shaft. The lode was divided into floors 65 ft. apart. From the extraction shaft galleries were driven at every floor, cross-cutting the lode entirely. Where these galleries reached the ore, narrow galleries were driven east and west, following the hanging-wall along all the sinuosities of the lode and determining its shape. From these galleries, again, cross-cuts were driven through the lode every 33 ft. After learning the shape of the lode, a drift was run in the hanging-wall at an average distance of 15 ft., and from the drift cross-cuts were driven toward the lode at regular intervals of 33 ft. The drift was equipped with track and used as an extraction gallery. It was found that the drift along the hanging-wall itself was not suited to use as an extraction gangway, as it required constant timbering. Both walls of the lode are slate and all workings in it required close timbering.

When the cross-cuts from the main hanging-wall gangway had reached the foot-wall, they were filled with rock carefully piled up, and new cross-cuts were then driven alongside the first ones. These were filled up, and again new ones were made and filled, and so on, until a slice of ore had been removed over the whole length and width of the deposit. All the galleries and cross-cuts had a uniform width of 6×6 ft., so that the height of the first slice removed was 6 ft. When this had been accomplished, the gallery along the hanging-wall also was filled with waste, except at the cross-cut connecting with the main gangway out in the hanging-wall. At this point a mill-hole was constructed by carefully building a rock wall around it.

When the first slice had been removed and the entire excavation filled with stone placed by hand, as above explained, a new hanging-wall drift was run directly over the first, following the wall, and from this a new series of cross-cuts was driven, starting immediately above those first driven and opposite the mill-holes in the drift on the wall. These new cross-cuts in turn were solidly filled with waste, and other cross-cuts were driven beside them and filled. While the first slice had been broken out of the solid ore body, the slice next above was under-cut over its entire area; in fact, it was resting on the filling. This made the blasting very much cheaper, so that while the contract price paid the miners for the first slice was 50 cents per ton, it was only 25 cents on the next slice above. The first cross-cuts driven through the solid ore body cost 76 cents per ton of ore, while the secondary cuts, which were only widening out the first ones, cost 42 cents per ton of ore. These figures show the advantage of having free sides for the working faces — an advantage not obtained in the pillar-and-stall method.

After the second slice has beeen removed, a third is taken out in exactly the same way, and so on, until the whole lift of 65 ft. has been removed. Several levels were being worked in this manner at one time, so that ore production was not limited. No new galleries nor cross-cuts were started until those adjoining had been properly filled in. As the slicing proceeded upward, the mill-holes were carried up by walling up the opening.

The rock used for filling was quartzite, quarried at the surface in the vicinity, and was delivered by tramway at the collars of two shafts sunk in the foot-wall for the purpose of delivering stone to the stopes. The rock is lowered in stone-boats, that for the fourth level being taken off at the third and trammed through the extraction gallery on that level and sent below through winzes, falling almost at the point that it is to be used. As the slicing proceeds, these winzes are filled up, while the ore chutes grow larger as stoping continues upward. The winzes are cut at the hanging-wall side, are 3 ft. square, and are securely timbered. No smalls are used in filling, only big stone being employed. A stope 65 ft. high does not settle more than 6 in. when properly filled.

Chapter XXVIII

HEAD-FRAMES

BY GEORGE SYDNEY BINCKLEY C. E.

An important element in the working equipment of a mining shaft is the Head-Frame, or Gallows Frame, as it is often called.

In almost every mining district a great range may be seen in the variety and size of the head-frames, from the simple tripod of the prospector or small "leaser" to the towering structures of timber or steel that rise over the deep shafts of great mines. Yet in spite of all this variation in magnitude and type, the principles underlying the correct design of the head-frame are the same, be the structure large or small.

Although the mathematical elements involved in the design of a head-frame are simple in the extreme, no important part of the equipment of the mine has been so slighted by the engineer as this. Hardly a mining camp can be found that does not exhibit several examples of large and expensive head-frames designed with a total disregard for the real nature of the stresses that they are intended to support.

The essentials in the design of a head-frame are:

1st. *Strength.* — This may best be assured by the distribution of the materials of construction along the lines of the stresses to be resisted.

2d. *Stability.* — This may best be secured by giving the structure as a whole the pyramidal form, by the avoidance of eccentricity of stresses in the structure, and the provision of sufficient area of base, within which the resultant of the working stresses shall fall.

3d. *Economy in construction.* — This will follow adherence to the rules which will be observed in securing strength and stability to the structure as outlined above, for the materials of construction distributed along the lines of stress will be employed in the resistance of those stresses with the highest degree of efficiency.

The working stresses in a head-frame (aside from the lateral stresses introduced in a vertical frame when self-dumping skips are employed, and those due to the weight of the skip in the case of an inclined shaft) are transmitted through the hoisting rope to the sheaves and their bearings, thence through the members of the supporting structure to the foundations.

The strain on the rope is the same on both sides of the sheave, and the direction of the strain is along the center line of the rope. The amount of weight on the sheave bearings will vary with the angle between the center of the rope leading to the hoist, and that from the sheave to the shaft. For example, if the rope led straight up from the shaft to the sheave, and straight down, vertically, from the sheave to the hoisting engine drum, the weight on the sheave bearings would be double that of the cage and its load, while if the angle between the center lines of the rope as described above is 90° — a right angle — the weight supported by the sheave bearings will be but 1.41 times the weight carried by the rope, and if the angle between the ropes be 120° the weight supported by the sheave bearings will be equal to the pull on the rope. In other words, the weight supported by the sheave bearings is equal to the pull on the rope (or weight of cage, load, and rope in a vertical shaft, or these weights multiplied by the sine of the angle of the shaft from the horizontal in the case of an inclined shaft) multiplied by twice the cosine of half the angle between the rope leading from the sheave to the hoisting engine and that from the sheave to the cage in the shaft.

The direction of the pressure on the sheave due to the tension on the rope, supporting the cage and its load on one end, and attached to the winding drum of the hoist on the other, is in the vertical plane only when the vertical angle of the rope leading to the hoist is the same as that leading from the sheave to the shaft. The condition described is seen in Fig. 115. In all other cases, the direction of the pressure transmitted from the ropes through the bearings of the sheaves to the supporting structure must be in other than a vertical plane.

In all cases, the *direction* of the pressure on the supporting structure lies midway between the center lines of the ropes leading from the sheave to the cage and from the sheave to the winding drum of the hoist.

A full understanding of these fundamental facts is necessary for the proper consideration of the problem of head-frame design, whether the structure is to be large or small, for light or heavy duty.

Reference to Fig. 116 will make clear to the reader the explanation given above relative to the direction of pressure due to the tension on the ropes. It will be seen in both instances given, that the resultant of these forces lies midway between the ropes, whether the shaft be vertical or inclined. Fig. 115 is a case where the inclined shaft is considered. Here, if the hoist is so set that the angle of the rope is the same on both sides of the sheave, the direction of the pressure on the structure (the resultant) will be vertical. It thus becomes entirely proper to make the back of the frame vertical, and to employ the front for the support of the skip tracks and the support of an ore bin as shown.

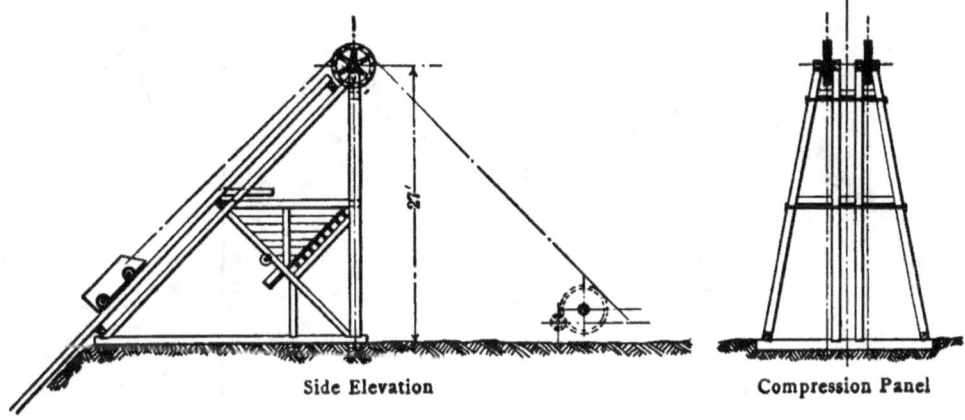

Fig. 115

This makes an extremely simple and economical design for a head-frame for an inclined shaft, and, while cheaper and more compact than the usual form, is quite as safe and stable as if the structure were extended far out toward the hoist. The hoist could in this case be set closer to the frame without any change in its design, but if set farther away, the back bent would have to be correspondingly inclined so as to fall in the line of the resultant.

In engineering, little is to be learned from success — much from failure. An error, to be corrected, must first be recognized. In order to make clear the errors so common in the ordinary type of head-frame so extensively employed, Figs. 117, 118, 119, 120, are introduced.

In Fig. 117, is seen a very common form of head-frame, ordinarily used where the shaft is of considerable depth and extraction heavy. On account of the complicated framing and the large number of bolts and ties called for in this design, the cost of this type of head-frame is generally heavy. Obviously, the assumption upon which this design is based is that the direction of the pressure to be resisted by the structure is in the vertical plane, along the center line of the rope leading down the shaft. This idea is strongly indicated by the position of the rope in the braced tower, and the position of the sheave.

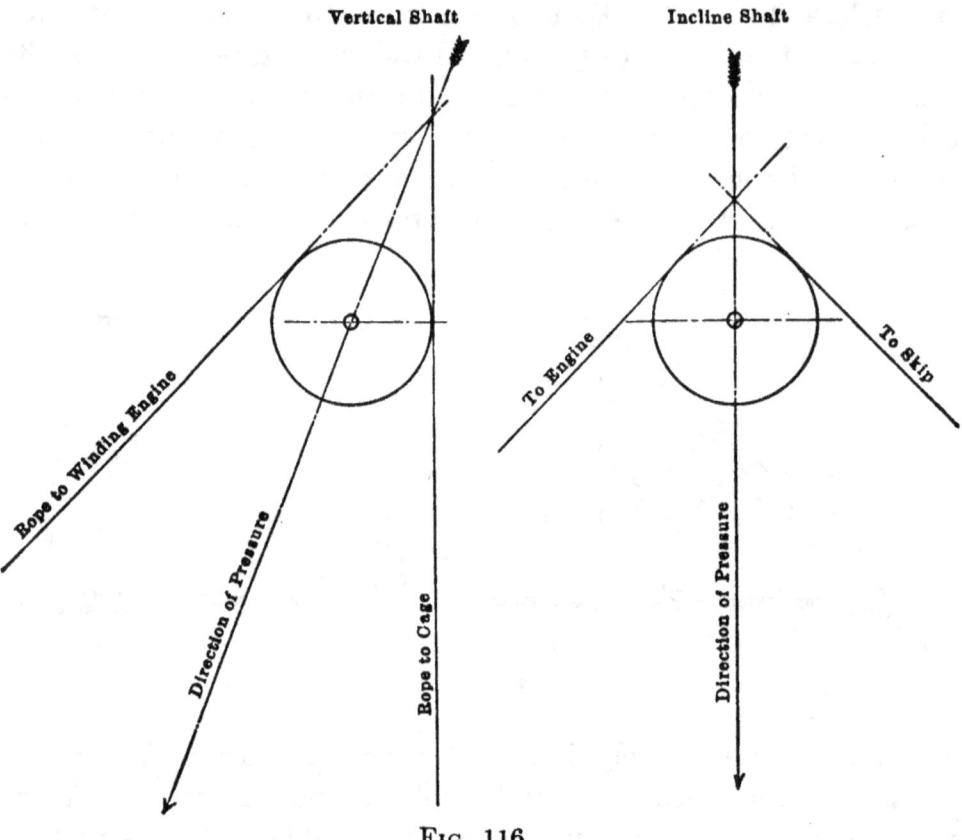

Fig. 116

The actual direction of the pressure due to the load on the rope is, in this diagram, very clearly shown, and it will be seen that the line of the pressure falls almost entirely outside of the elaborately framed and braced four-post tower, and is taken almost altogether by the back bracing. As a matter of fact, the four-post tower could have substituted for it a plain vertical bent without sacrifice of strength or stability.

In Fig. 118 is shown the familiar simple tripod. In this case it will be observed that the direction of the resultant is well within the simple members of the frame, and that this

HEAD-FRAMES

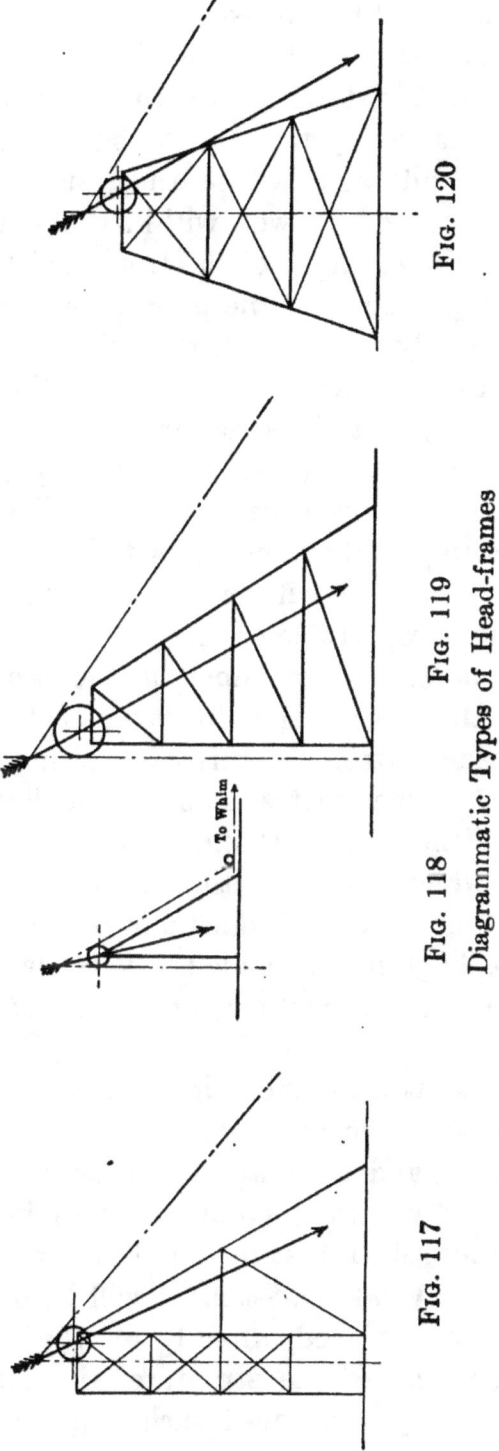

Fig. 117 Fig. 118 Fig. 119 Fig. 120
Diagrammatic Types of Head-frames

elemental form of head-frame is far more correct in design than the ambitious and complicated "four-post" type.

A type of the head-frame used to a considerable extent in many mining districts is shown in Fig. 119. This type is a great improvement in stability over the "four-post" design, yet its complicated bracing system should be avoided as expensive and irrational. While in this type the resultant of the tension on the ropes generally falls well within the structure, a proper relation between the strength of the front and back members of the frame is rarely seen, and the practice of placing the sheave bearings on relatively long caps connecting the front and back panels introduces a sequence of strains in the bracing system that extends throughout the structure to the foundation. This fact makes necessary the very heavy bracing system always found in this type of head-frame, as its duty is not alone to secure stability to the structure, but to transmit part of the working strains to the foundation — a duty with which the bracing system of a head-frame is not properly charged.

Fig. 120 is a striking example of the scant consideration so often given the nature and direction of the stresses to be resisted by a head-frame. In this case, although the structure is apparently one of the greatest stability, the line of the resultant of the tension on the ropes falls almost entirely outside the frame, the stability of which depends altogether on its own dead weight and that of the foundations to which it is anchored. The example from which this diagram was made is a very elaborate steel structure, and is illuminating in the completeness of its contempt for rational design.

The writer has made an effort, in the typical design for a large steel head-frame shown in Figs. 121 and 122, to avoid those errors pointed out above, and to conform as nearly as possible to both the theoretical and practical requirements of head-frame design. In this design (published first in the Mining and Scientific Press of San Francisco several years ago) it will be observed that the sheave bearings are placed directly on the end of the main compression members of the structure, the intermediate bearings being carried by a beam of such depth and stiffness that deflection under load will be negligible. The two main columns, being brought as close together as possible at their heads, are inclined laterally, giving a relatively wide base and consequent stability. The position in which the sheave bearings are placed insures the direct transmission of the working stresses through the main compression members to the foundation without eccentricity, hence the efficiency of the metal in the columns is maximum.

On account of the fact that the working resultant of the tension on the ropes lies practically within the compression member itself,

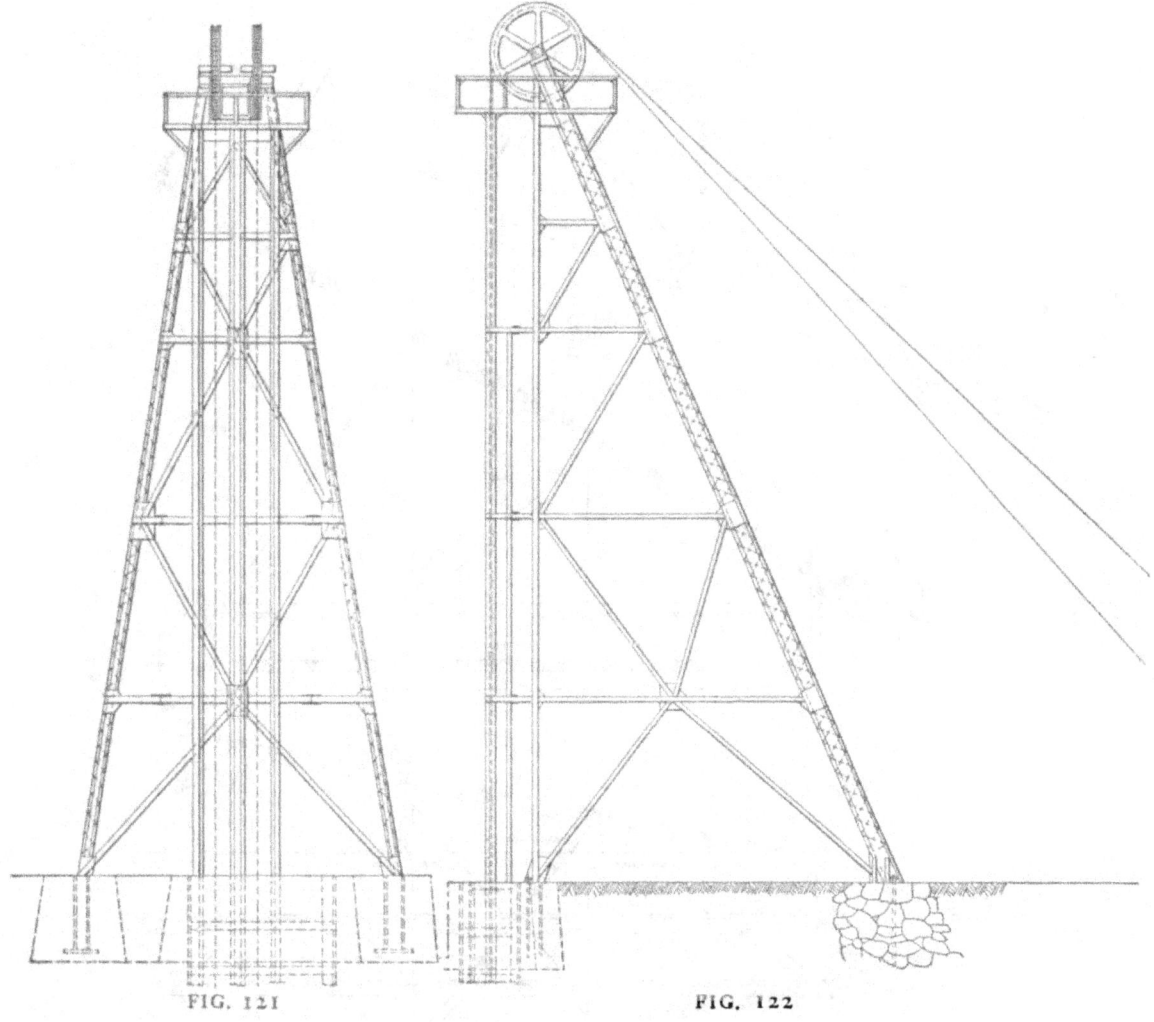

FIG. 121 FIG. 122

the function of the forward or shaft side of the frame is merely to hold the heavy back panel in position, and give the necessary stability to the structure. This bracing may be very light, for no working strains whatever are transmitted through the bracing system, except those insignificant strains that may arise from the dumping of skips if these are used.

The design exemplified in Fig. 121 probably represents the ultimate form of greatest economy of material in the construction of a large steel head-frame for a vertical shaft.

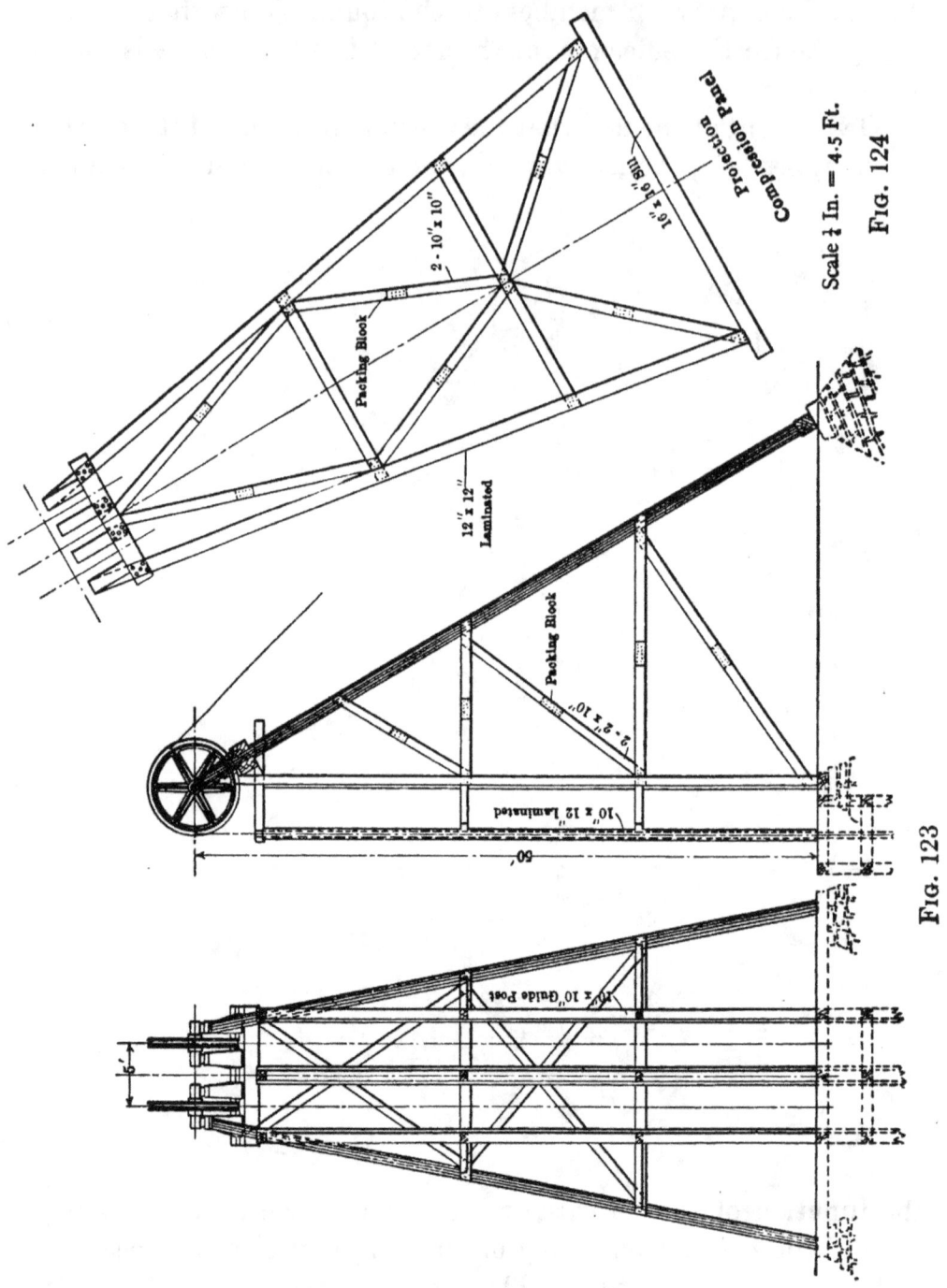

An adaptation of the same design is seen in Figs. 123 and 124. This is a typical design for a fifty-foot head-frame of wood construction, and embodies the same characteristics as those of the steel head-frame described above. An unconventional method of construction, however, is suggested in this design, which the writer believes will be found of value.

It will be seen that the timbers, instead of being solid, are of a laminated structure, built up of two-inch plank spiked together in such a way that the bracing system is built in with the main compression members, requiring for the most part no rods, bolts, or ties, to secure a very solid and strong frame.

The advantages offered by the proposed method of construction are several. Long timbers are not necessary, as members of any length may be built up out of comparatively short plank. Such timber or plank as would be needed for the construction of even a very heavy head-frame of this type can be much more readily obtained than long dimension timber, and at much lower cost. It is also true that such a frame as that shown in Figs. 123 and 124 will be stronger than one framed of heavy timber, while it can be successfully built with less highly skilled labor.

It is obvious that many variations in the detail of construction and design will present themselves to the engineer during the working out of field drawings for the actual construction of such a head-frame, but the general principle shown can be followed with full confidence in the strength and permanency of the completed structure, whether it be large or small.

An earlier design by the writer is seen in Fig. 125, which is an illustration of the High Ore head-frame, Butte, Montana. This, and another of equal height to the center of the sheaves for the Diamond Shaft, was designed by the writer in 1898, and although not so advanced as that shown in Fig. 121 embodies most of the principles elucidated above. It is one hundred feet high to the center of the sheaves, and carries a live load on the ropes of about sixty thousand pounds. The hoisting engine (also of the writer's design) is of the horizontal, double, direct-acting type, with cylinders 30 inches diameter by six feet stroke, working under a steam pressure of 140 pounds to the square inch. The duty of this head-frame is obviously very heavy, yet on account of the fact that the sheave bearings are carried directly on the main members, and the pyramidal form of the structure, it is perfectly stable and free from any vibrations even under full load moving at high speed. The relatively small base of this head-frame makes its stability still more noteworthy, and demonstrates the correctness of the principles followed in its design.

Two other steel head-frames in Butte, designed by the writer for the Boston & Montana Co. and erected on the Mountain View and the Leonard shafts of that company, were eighty feet to the center of the sheaves, and followed in general the same prin-

Fig. 125

ciples of design employed in the High Ore and Diamond head-frames. Yet, although perfectly stable and satisfactory in service, the writer is now forced to consider them relatively crude, especially as regards economy of construction, when compared with those shown in Figs. 121 and 123.

The head-frame on an important mine is now such a serious element in the equipment that it should receive the intelligent attention of the designing engineer, and be no longer left to the tender mercies of the boss carpenter, or the well-meant but often unfortunate attentions of the mine foreman.

INDEX

	PAGE
Abuses of square-set system	193
Air currents, effect of, on timbers	2, 3, 4, 5, 6
Alma mine, Cal., shaft	75, 77
Angle braces in square-sets	191
for side pressure	191
Argonaut Mine, Cal.	55, 75
Automatic dumping devices at shafts	64, 72
winzes	184
skips	128, 129
Bacon, D. H.	166
Bad ground	19, 28
Bailing by electrically-operated skips	51
Bailing by skips and tanks	125, 127, 130, 132
Barrier Range, Australia	213
Bearers in shafts	105, 162
Beaumont, E. K.	213
Behr, Hans C.	126
Bevel in tramming shaft sets	75
Big Indian mine, Mont., open cut at	135
Binckley, George S.	255
Blasts fired electrically	118
Blasting holes in series	118, 174
Block caving at Pewabic mine	172
Block holing at Homestake	240
Blocking out ground for caving system	174
Block system of stoping at Broken Hill	213, 223, 224
Homestake	242
Blue lead of California	38
Bolts, hanging	58
Breast boards in working through running ground	29, 32
shaft sinking	99
Breasting posts and caps	38
Bridge, the use of in drifting	24, 29
shaft sinking	99
Broken Hill Mines, N. S. W. Australia	193
block stoping at	213, 223, 224
open-cut mining at	226
stoping by slicing at	222
underground open-cut at	218

INDEX

	PAGE
Brush treatment of timber	9, 14
Bucket-dumping devices	64, 72
use of in sinking	58
with valve	127
Building of chutes	170
Building up to secure dump	55
Bulkhead in stopes at Broken Hill	219
Bumper for crosshead	61
skips	71
Cable mat to protect shaft timbers	112
Cages in shafts	53, 119, 120
skips	120
service, in shafts	50
Caledonia mine, South Dakota	236
Calico District, Cal., large stopes in	195
California drift mines, method of working	38, 41
Capacity of shafts	50
bailing skips	51
Caps and breasting posts	38
replaced in position	31
Cap sills in square-sets	207, 216
Carbon dioxide, effect of on timbers	7
Care in placing posts in drifts	159
Cave in Caledonia mine	236
mine at Angels, Cal.	195
Caved ground working at Homestake	238
Caving system at Iron Mountain	172
Centering drift sets	18
Cerro Prieto mine, Sonora., Mex., mining at	154
China, great shafts in	49, 73
Churn drill or jumper	141
depth of holes drilled by	143
Chutes for passing timbers underground	203
in square-sets	167, 211
construction of	170, 211, 214, 235
Clevice for handling timbers at shafts	87
Cobalt district, Ontario	62
small shafts in	63
Collar of shafts	44, 56
concrete at	44
Combination inclined and vertical shafts	102, 103
of round and square timbers in sets	200
Comparative cost of timber treatments	13
Compartments, size of shaft	54
Comstock lode, early mining on	187
introduction of square-sets on	188

INDEX

	PAGE
Comstock lode, shafts on	73, 78
skips at	128
Concrete at collar of shaft	44
Connecting levels by raising	161
in stoping	184
Construction of square-sets	201
Continuous raises at Homestake	233
Cornish pumps	51
Correcting errors in placing shaft timbers	92
Cost of stoping at Homestake	245
timber treatment	13
Cribbed shaft, how built	48
Cribs	46, 48
in square-sets	193, 221
surrounding shafts in swelling ground	109
"Creeps"	216
Cross-head in shaft sinking	58
how made	59
bumper for	61
Cut holes, best place for in blasting	118
in Hoatson shaft, Bisbee, Ariz.	118
Cut stations when sinking	120
Cutting and timbering stations at shafts	119
stations, methods of	120
Cylinder treatment of timbers	13
Danger in overwinding	72
of fire in timbered stopes	217
Deadwood-Terra mine, stoping at	238
Deep-level shafts on Witwatersrand	104
Deidesheimer, Philip	188
Del Mar, Algernon	62
Delprat, D. G.	253
Details of framing shaft timbers	48, 74, 76, 77, 79
square-set system	192
Detroit Copper Co., Morenci, Ariz.	45
Diamond mines, South Africa, mining at	247
Different styles of drift timbering	17, 18, 183, 227
head-frames	255
Disadvantages of underhand stoping	149
Direction of strains in head-frames	256
Distance piece for guides in shafts	82
Divider, the	80, 83
Dogs, iron, in shafts	78
Douglas spruce	2, 3, 4, 5, 6
Drainage of shafts	51
by bailing	51, 126, 132

	PAGE
Drains in drifts, necessity for	17
Draining wet, running ground	28
Drifts, drains in	17
Drift sets, how centered	18
lining up	25
mines, timbering of	38, 41, 159
timbering at Broken Hill	227
timbers crushed by pressure	108
Drifting and drift sets	16, 17, 18, 183, 228
with false set	24
Drill holes in shaft sinking	117
"springing" for heavy charges	143
Drilling with churn drill	142
Driving lagging	18, 24, 28, 29
Dumping devices at collar of shafts	64, 72, 128
winzes	184
Eagle-Shawmut mine, Cal.	155
Early mining at the Broken Hill mines	214
Comstock mines	187
Homestake	236
Economy in head-frame construction	255
Effect of moisture on timber	2, 3, 4, 5, 6
air currents on timber	2, 3, 4, 5, 6
carbon dioxide on timber	7
Elkhorn mine, Mont., remarkable timbering in	188
Ellison shaft, Homestake	229
Empire mines, Grass Valley, Cal.	2
Evolution of square-set timbering	188
Extension tracks for shaft sinking	113
wheels on skips	71
Results derived from experiments in treating timber	13
Face boards in working through running ground	29, 32
sinking shafts	99
False set, drifting with	24
Fenders to protect shaft timbers	111
Filling at Broken Hill	216
Homestake	140
in stopes	140, 168, 214
for stopes, how obtained	160, 181, 254
waste used for	152
Finlay, J. R.	154, 156
Fire, danger of in square-sets	217
Firing blasts by electricity	118
Flat veins, method of stoping	158
"Flying Fox" at Broken Hill	226

INDEX

	PAGE
Foot blocks in drift sets	29
Forepoling in shaft sinking	99
Framing shaft timbers	47, 73, 75
for top pressure	198, 199
square-set timbers	197
timbers for side pressure	198
Fungus growth on timbers in mines	2, 3, 4, 5, 6
Gallows frames	255
Government experiments in timber treatment	8, 15
Grizzly in open cut	134
ore pass	137, 232
Guides in shafts	59, 61
Gwin mine, California	178, 183
shaft	55
Half hitch and timber hitch	89
Hammer drill, pneumatic	170, 240
Hampton, E.	5
Handling powder underground	123
timbers in raises	180
shafts	58, 87
Hanging bolts in shafts	57, 78
Haste in bad ground a mistake	28
Hayward, Alvinza	205
system of square-sets	205
Head-frame on framework	55
rational design of	255
steel	53, 263
Height of shaft stations	122, 229
Hewn timbers	48
High ore mine, Butte, Mont., steel head-frame at	264
Homestake, block system of stoping at	242
continuous raises at	233
early mining at	236
filling stopes at	240
mills	235
mine, South Dakota	120, 123, 136, 229, 238, 242
mining at	229
recovering caved ore at	236
shaft stations at	229
stoping without timbers	238
Huddlestone, Thos.	157
Hydraulic jack, usefulness of in mine	31, 58
Improper head-frame design	255
Inclined shafts	55, 69, 71, 76, 93, 103, 107, 113, 115, 124, 129

	PAGE
Inclined shafts, bucket dumping devices at	64, 72
in hard rock	115
Inverted rails for track extension	114
Iron dogs in shaft timbers	78
Iron Mountain, caving system at	172
Jack-knifing of sets	22
screw in mines	31, 58
Joint for post and sill	23
Jumper or churn drill	141
Kennedy east shaft	55, 99, 101
Kimberley diamond mines, mining at	247
Kinds of timber used in mining	1
Knots, some useful	88, 89
Ladders in raises	163
Lagging, driving	18, 24, 29
in square-sets	214, 240
on floors of stopes	242
Laminated construction of head-frames	263
Large ore bodies, mining by open-cut method	133, 144, 227
stoping	187, 207, 209, 212, 238, 242, 254
tonnage hoisted through shafts	50
Lateral stresses in head-frames	256
Leggett, Thomas H.	104
Levels connected by raises	161
connecting, method of	184
Life of timbers underground	2, 3, 4, 5, 6, 206, 214, 216
Limitations of square-set system in mining	195
Lining up drift sets	25
shaft sets	93
Location of shaft with reference to subsequent operations	43
Long wall system of mining	160
Lord, Elliott	187
Machine shops underground	122
Main gangways in country rock	179
Makeshift methods unadvisable	57
Making a cross-head	59
Method of cutting stations at shafts	120, 123, 124
framing drift sets	17, 18, 183, 226
square-sets	197, 209
Method of mining at Broken Hill	193, 212, 223, 225
Homestake	233, 239, 242
Utica-Stickle	208
Yellow Aster	137

Method of mining at Zaruma, S. A.	156
by Longwall system	160
in open cuts	133, 144, 226
obtaining filling from walls	182
stoping	148, 154
flat veins	185
in weak walls	166
slices	156, 222
swelling ground	178
without timbers	172, 238, 253
timbering at Gwin mine	183
Utica mine	208
Wildman mine	211
working drift mines in California	38, 41
Mexico, great shafts in	49, 73
Mill holes in open cuts	134, 137
Mine drainage	17, 28, 125, 127, 130, 132
Mine drainage and pumps	126
Mine models, usefulness of	243
timbering, methods of	40
Mining at Broken Hill, N. S. W.	212
Cerro Prieto, Mex.	154
the Homestake	229
Kimberley	247
Angels, Cal.	208
Yellow Aster	137
by caving system	172
Glory Hole method	133, 136, 143
Longwall method	160
Minnesota iron mine, stoping in	166
Mistakes in use of square-sets	193
Modifications of square-set system	201, 203
Moisture, effect of on timbers	2, 3, 4, 5, 6
Mother lode of California	19, 55, 75, 106, 178, 195, 209
Mount Morgan mine, Australia	139
Lyell mine, Tasmania	147
Necessity for drains in drifts	17
New Almaden mine, Cal.	5
Norris, R. V.	34
Oak, use of in mines	3
O'Brien, W. S.	239
Open cut Big Indian mine	135
Broken Hill mine	226
churn drill in	141, 144
Homestake mine	235

	PAGE
Open cut Yellow Aster mine	137
Wasp No. 2 mine, South Dakota	158
method of mining	133, 144
mill hole in	134
steam shovel in	144
underground at Broken Hill	218
stope	151
tank treatment of timber	10, 13
Ophir mine, Comstock lode	187, 194
Ore chutes in square-sets	176, 211
Ore-bins beneath levels	120
Oregon pine	1, 2, 3, 4, 5, 6, 216
Ore pockets under levels	120
Overhand stoping	150
stope carried to surface	151
Overwinding, danger in	72
Oneida mine, Cal.	5, 21, 55, 101
Patton, W. H.	193, 214
Permanent hoisting plant	52
Pewabic mine, caving system at	172
Pine, varieties of, used in mines	2, 3, 4, 5, 6
Placing shaft timbers in position	89, 95
sills in square-sets	202
stope timbers in position	201
Platforms in drift sets, temporary	26
shafts, temporary	91
stopes, temporary	203
Pneumatic-hammer drill	170, 240
Policy for preservation of timber	14
Position and direction of drill holes in shaft sinking	117
of temporary hoisting plant	52
Possibilities of timber treatment	11
Post and sill joint	23
Posts, care necessary in placing	159
perpendicular to roof	21
Powder underground	124
Premium system	46
Preservative treatment of timber	8, 14
Principles underlying methods of timbering	16
Prospectors as engineers	43
Prospecting shafts	43
Protection of shaft timbers	111
Pumps, Cornish	51
Raises, continuous	233
divided	161

INDEX

	PAGE
Raises for connection of levels	161
ladders in	163
how made	161
ventilation of	165
Raising, progress of at Homestake	234
Redwood as mine timber	2, 3, 4, 5, 6
Reinforcing square-sets	193
Remarkable stope in Elkhorn mine, Mont.	188
Repairing shafts	116
Resultant strains on sheave in head-frame	256, 259
Results of experiments in treatment of timber	13
Robbins, Frank	69
Ross, John, Jr.	2
Ross, Gilbert McM	4
Rowlands, Richard	29
Running ground	28
driving through	29
sinking through	97
Saddle wedges, use of	31
Sagging caps, how replaced	31
Scarcity of timber, effect of on mining	154
Selection of timber	47
Self-dumping skips	128, 129
Service cage in shaft	51
Shaft building by raising	161, 163
compartments, size of	54
Ellison, at Homestake	229
repairing	116
sinking, drill holes in	117
timbers, placing in position	89, 93
protection of	111
Wildman, Cal.	107
Shafts, bucket dumping devices at	64, 72, 128
combination vertical and inclined	102, 103
concrete at	44
cribbed	46, 48
guides in	59, 61
handling timbers in	85
large tonnage hoisted through	50
location, kind and size of	43
on Comstock lode	73
size and division of	49
without timbers	46, 47
working	43
Side pressure, angle-braces for	191
framing timbers for	198

	PAGE
Sills, employment of in drifts	22
stopes	190, 201, 207, 211, 214
placing in square-sets	202
Sinking shafts, extension tracks for	113
ladders for	115
through running ground	97
with cross-head and bucket	58
Size and division of shaft compartments	54
of timbers determined by experience	20
stopes depend on ground	203
Skips, automatic dumping	128, 129
bailing with	125, 127, 130, 132
in inclined shafts	58
versus Cages	120
with valves	127
Slicing at Broken Hill	221
Zoruma	156
Slides for timber in square-sets	203
Sloping stopes at Broken Hill	222
Some useful knots	88, 89
South Eureka mine, handling timbers at	180
Spiked plank in drift sets	17, 18
Spliced wall plate	81, 84
Sprags, defined	22
in square-sets	22, 211
use of	22, 211
Springing drill holes for blasting	143
at Wasp No. 2 mine	158
Square-sets, construction of in stopes	201
Hayward system	205
Square-set system, abuses of	193
invention of	188
limitations of	195
method of framing timbers for	197
modifications of	201
wall plates in	191
Starr, Geo. W.	2
Stations at shafts	119
Homestake shafts	229
height of, at shafts	122, 220
inclined shafts	124
vertical shafts	123
Steam shovel at Bingham, Utah	147
Ely, Nev.	146
Granby, B. C.	145
in open cuts	144
Mesabi Range	145

	PAGE
Steel head-frame, High Ore mine, Butte, Mont.	264
Old Dominion mine, Globe, Ariz.	53
mine timbers	34, 37
Stope, caving of at Angels, Cal.	195
chutes in	167
connecting levels in	184
"creeps" in	216
filling in	168
size of	194, 195
superficial area of	194, 195
without timbers	238, 253
Stoping at Broken Hill	216
Homestake, cost of	245
Utica mine	208
by block system, Broken Hill	212
at Homestake	242
in diamond mines, South Africa	247
swelling ground	178
walls for filling	181
weak walls	166
methods	148, 160
large ore bodies	187
early difficulties	187
overhand	150
underhand	148, 149
without timbers at Homestake	233
in Spain	253
Strain on hoisting rope	256, 257, 258, 261
Stulls, use of in small shafts	62
Sump tanks under levels	125
Swelling ground	19, 106, 180
on mother lode	106, 180, 183
sinking through	109
stoping in	178
Tanks, bailing	125, 127, 130, 132
Tanks under levels	125
Template, the	85
how made	85
Temporary hoisting plant	52
platforms in drift sets	26
stopes	203
The straight-edge	94
bowline	89
cable mat to protect shaft timbers	112
cross-head, how made	58, 60
divider	80, 83

	PAGE
The template	85
timber hitch	89
truck	85
Timber chutes in stopes	203
crushed by pressure	108
framing machine	199
gang	27
head-frames	262
selection of	47
truck, the	85
effect of air currents on	1, 2, 3, 4, 5, 6, 216
handling at shafts	85
hitch	89
preservation of	8
treatment of	9
placing in position in shafts	89, 93
stopes	201
used in mines	1, 2, 3, 4, 5, 6, 214
Timbering at Broken Hill	212, 225
Gwin mine	178, 183
Homestake	229
Utica mine	208
drift mines	38, 41
stations at shafts	119
Top pressure, framing for	198
Topography about shaft important	53, 55
Treatment of timber in open tanks	10
cylinders	13
by brush	9
possibilities of	11
Treadwell mine, Alaska	244
Tuffs, large stopes in	195
Types of head-frame construction	259
stations at inclined shafts	124
vertical shaft	123
Utica-Stickle mine, Angels, Cal.	55, 86, 178, 208
handling timbers at	86
stoping at	208
Usefulness of mine models	243
square-set in extracting large ore bodies	191
hydraulic jack in mines	31
Underground machine shops	122
open cut, Broken Hill	218
working powder in	123, 124
Underhand stoping	149
Underlying principles in timbering	16

INDEX

	PAGE
Unusual methods of stoping	154
Uren, Charles	155
Valve buckets and skips	127
Varieties of timber used in mines	1, 2, 3, 4, 5, 6, 214
Ventilation of raises important	165
at Homestake	234
Vertical shafts, bucket dumping	64
devices at	64, 72
placing timber in	89
Wall plates in shafts	81
square-sets	191
Walls, weak, stoping in	166
Wasp No. 2 mine, mining method at	158
Waste for filling	152
Weber, Frank F.	5
Wedges, importance of	199
must be kept tight	47
Wildman mine, Cal.	4, 107
Williams, Gardner F.	247
Winzes, dumping devices at	184
Witwatersrand, deep-level shafts	104
Working caved ground at the Homestake	237
shafts	43
stresses in head-frames	256
Yates, Bruce C.	244
Yellow Aster, mining at	137
square-sets in	195
Yellow pine	2, 3, 4, 5, 6
Zaruma, Ecuador, S. A., mining at	156